SERGO ORDZHONIKIDZE RUSSIAN STATE UNIVERSITY FOR GEOLOGICAL PROSPECTING
FEDERAL SCIENTIFIC CENTRE VIEV
SHEMYAKIN-OVCHINNIKOV INSTITUTE OF BIOORGANIC CHEMISTRY, RUSSIAN ACADEMY OF SCIENCES

PROCESSING TECHNOLOGY OF UNIQUE RUSSIAN CARBON-BEARING ROCKS – SHUNGITE

VLADIMIR A. RAFIENKO
VLADISLAV V. BELIMENKO

Translated (from Russian)
by Daria V. Dementieva

I0145898

ACADEMUS
Publishing

Academus Publishing
2019

Academus Publishing, Inc.

1999 S, Bascom Avenue, Suite 700 Campbell CA 95008
Website: www.academuspublishing.com
E-mail: info@academuspublishing.com

The right of Vladimir A. Rafienko, PhD in Mining and Engineering,
Federal State Budgetary Institution of Higher Professional Education
"Sergo Ordzhonikidze Russian State University for Geological Prospecting";
Vladislav V. Belimenko, PhD in Biology,
Federal State Budget Scientific Institution "Federal Scientific Centre VIEV"
is identified as author of this work.

Translator: Darya V. Dementyeva, PhD in Bioorganic Chemistry,
Shemyakin-Ovchinnikov Institute of Bioorganic Chemistry,
Russian Academy of Sciences

For e-mail correspondence:
Dementyeva D.V. — dasha160370@icloud.com, npp-f@yandex.ru;
ORCID: https://orcid.org/0000-0001-9123-9866, Researcher ID: Y-7312-2018

Rafienko V.A. — vrafienko@mail.ru
ORCID: https://orcid.org/0000-0002-4349-5539, Researcher ID: Y-8759-2018

Belimenko V.V. — vlad_belimenko@mail.ru
ORCID: https://orcid.org/0000-0001-8871-7863, Researcher ID: A-7954-2017

Reviewers:
Ilias, Nicolae
Professor, PhD Engineering
Member of the Academy for Technical Sciences of Romania

Volfson, Iosif
PhD in Mineralogy and Geochemistry
Academic Secretary of Russian Geological Society

ISBN 10: 1 4946 0012 9
ISBN 13: 978 1 4946 0012 9
DOI 10.31519/1192

Major problems of processing technology of shungite rocks are illustrated in the book; ultimately new integrated technological classification of all shungite rock types, which has allowed to expand their application sphere by far, is given. On the basis of proposed classification, the technological evaluation methodology has been carried out and its theoretical substantiation is given. It is shown that sulfides and carbonates, contained in the shungite rock, should be exposed to chemical leaching process. For the first time, physico-chemical parameters of leaching process technology of sulfides have been established; deformation mechanism of sulfides during their drying has been figured out. It is proved that by the charring way, paroptesis and autoclave processing the series of valuable products can be obtained from the shungite rocks: liquid glass, sodium carbide, calcium chloride, metallic calcium, lime, cement. The book can cause significant scientific and practical interest for wide range of specialists, working in the sphere of natural resource enrichment that allows to recommend it for scientists and technical school teachers. It can also work as a valuable guidebook for engineers-technologists, projectors and employees of spheres allied with enrichment, and, thanks to availability and systematics narration of scientific research and technological bases of shungite raw processing — for technical students of mining specializations.

INTRODUCTION

The following book is devoted to the theoretical and practical technology evaluation of physico-chemical properties of shungite rocks — the new raw material as the substitute to such row of traditional materials as: chark, silica white, technical carbon and other components. Currently, shungite rocks have been widely used in ferrous and non-ferrous metallurgy, construction industry as well as in rubber technical goods production. The shungite rocks have been found for applications as mineral fertilizers, pigment for earth colors and other spheres.

Shungite rocks of Zazhogino deposit in Karelia is the unique natural multimineral formation, consisting of two components — carbonaceous matter, mineralogically close to graphite, and micro-crystal earth silicon. These rocks possess high strength, density, chemical stability and conductivity. The shungite rocks are very active in redox processes, especially at high temperatures. This gives an opportunity of their wide application in the iron metallurgy and the metallurgy of non-ferrous metals. They are also used while fusion of phosphorus-containing rocks to produce yellow phosphorus.

The shungite rocks — is a valuable mineral that can be widely used in different industry branches. In near years, it is planned to start construction of the houses with shielding properties that will reduce a harmful impact of electromagnetic waves of technogenic origin on human organism. This is extremely important in the house construction nearby airports, flag and transmission cell phone towers. These questions settlement can be reached thanks to regular shungite rocks addition into the construction materials.

The shungite rocks were formed from neptunic ones 2 billions years ago. Today, there is no consolidated view on their occurrence mechanism. A mechanism of physico-chemical interaction of shungite rocks with other mineral formations also has not been studied. The majority of specialists consider shungite rocks to be formed from sediment beds of lakes, seas and ancient sapropels. Nowadays, Zazhogino deposit — the largest and rather explored deposit of shungite rocks. It is in Karelia in the area of Onega Bay of the White Sea.

From ancient times, the peasants used shungite rocks as a remedy, a black ink, a local fuel and a decorative stone. Their specific and rather multi-side practical application has given scientists an opportunity to find a series of their new more promising properties. Thus, it has been observed that shungite rocks can be used as a filler in rubber technical goods, including automobile tires. The tires with shungite rock addition are more temperature-, cold- and wear-resistant. As well as they have become promising in the construction of warm floors with shielding properties in the residential buildings.

Being a good adsorbent, the shungite rocks have been already applied for sewage purification today. It is known about the practice of their use in the sewage managing facilities in Moscow. With the help of these rocks, flow waters from Moscow Ring Automobile Highway are being purified from oil and other organic compounds.

Taking into account the perspective of shungite goods application with various classes of coarseness — from finely-dispersed (5–70 microns) to coarse-grained (40–100 microns) — in Moscow and Moscow region, Moscow Government has adopted the decision on the construction of experimental-industrial enterprise for the production of shungite goods at Lianozovo district (OJSC "MKK-Holding", Moscow) and allocated the money for that purpose.

Simultaneously, large complex of research on further investigation of physico-chemical properties of shungite goods is being held and the sphere of their application in different industry branches and household are being specified.

Current manuscript is not claiming exhaustive generalization and research of all the problem of shungite goods application. It presents one of the first attempts to clarify all complex redox features, occurring while shungite products usage, especially in rubber technical goods, construction industry as well as in pyrometallurgical fusions in ferrous and non-ferrous metallurgy, and to find on the obtained results basis the new economically substantiated spheres of its application.

Sufficiently full settlement of the problem is possible only in longtime industrial usage as far as shungite products, selected at different deposit points, vary a little in their chemical composition and physico-chemical properties. But an impact of separate varieties has not been yet fully substantiated by practice, as well as there has not been yet specified the chemical composition of shungite rocks, the content of carbonaceous matter, earth silicon and other constituents in the rocks compositions that can be effectively used in different industrial branches. For practical usage in different branches, one should, first of all, be aware of the content of carbonaceous matter and earth silicon in shungite rock.

The current monograph analyses the earlier research on shungite rocks from the point of their possible usage in various industrial branches.

It is shown that major jobs in this direction were fulfilled mostly on rocks with average carbonaceous mater content of 25–40 %, while there is a big quantity of various rocks in the deposit with carbonaceous matter content from 5 to 20 %; together, there are rocks, where carbonaceous matter concentration constitutes 40–98 %. A little attention was paid to these rock groups

4

at technological evaluation earlier. That is why we have put a purpose — to work out the principally new technological classification of all shungite rock varieties with account of their chemical and mineral composition.

The obtained classifying complex technological scheme of shungite rocks is arranged by us with account of genetic features of all their existing varieties at the deposit and is demonstrated in triangle diagram.

On the basis of the elaborated complex classification and also the balancing theoretical diagram and the ratio of all existing shungite rocks, we have worked out methodology of technological evaluation of high-grade as well as low-grade shungite rocks. It is established that among low-grade shungite (shungitous) rocks one can specify the following kinds: silicate, aluminosilicate, silicate sodium, aluminosilicate carbon-bearing, carbon-bearing.

It has been proved that every kind of shungitous rocks can have only its own, based on its physico-chemical properties, conditions of application. Such assessment of these rocks has become possible due to the made classification. For each considered rock kind an own specific application is distinguished. Such complex approach widens a little bit the application area of such rocks and gives a chance to come up with new ways of their further perfection by further additional beneficiation.

Due to set classification, according to developed technological evaluation of shungite rocks one can quickly state their kind and potential application sphere.

Herein, the properties and specificities of physico-chemical behavior of all rocks kinds have been considered while their usage in various branches of industry.

It was figured out with the help of differential-scanning calorimetry that physico-chemical properties of different kinds of shungite rocks essentially vary. Thus, massive and brecciated rocks have different physico-chemical properties despite the same chemical composition. Along this, in such rocks kinds, the sizes of crystals of cherty minerals in the major matrix of carbonaceous matter range by far. So, in the massive rocks they constitute 1–2 microns, in the brecciated — 5–10 microns. It is shown that crystal lattice structure in the massive and brecciated rocks is also different. It says that, in the process of industrial usage of various massive and brecciated rocks, their activity mechanism will differ. These questions should be, certainly. taken into account while their usage in different industrial branches.

The mechanism of acid-forming characteristics in shungite rocks has been elucidated and the ways of their elimination in the industry processing flow are put forward.

For the first time, the elimination mechanism of sulphides from shungite rocks has been developed. The essence of this mechanism is that original rock with 2–4 % sulfide content acidifies initially by air oxygen in the apparatus of Pauchuk type, and then is exposed to leachate such as sulfuric acid medium at pH 3.5–3.8. It is shown that sulfides can be deleted from shungite rock in alkaline medium at pH values of more than 8.

It has been established that at pH from 4 to 9, the leaching of sulfides does not proceed, and vise versa, in fluid weak sulfurous and alkaline mediums, the sulfide buildup proceeds, while cellular pyrite is formed at 4–6 pH and pyrite — at 6–8 pH. This should be taken in consideration while pursuing the process of leaching; the developed mechanism is based on that matter. On the basis of obtained data, the technological scheme of leaching of sulfides from all shungite rock kinds has been developed

According to this principle, analogously, we have developed the technology of boart leaching in slightly acidic medium at pH from 4 to 6 with the help of hydrochloric acid. The resulting in the process of exposition, the calcium chloride solution is needed to be separated from solid body. Following such scheme, one can completely eliminate all of the boarts from shungite rocks.

Shungite rocks acquire novel properties after sulfide and boart leaching. This allows to widen the area of their application. Also, the technology of elimination of boarts from carbonate shungite rocks has been carried out.

Novel mechanism of shungite rocks drying-out has been developed that has allowed to substantiate the temperature and technological parameters of this process. Drying-out temperature should be no more than 300 °C.

Novel technology of closed water cycle with additional elimination of sulfurous gases and with acid waters neutralization has been theoretically stated.

With the purpose of pollution reduction, in addition to major project, the technology of air purification from sour gases, such as SO_2 and SO_3, and finely-dispersed solid shungite particles with coarseness from 0.1 to 5 microns has been offered. This technology provides with 100 % capture of harmful gases and finely-dispersed particles. To embody the given technology it has been proposed to organize the closed water cycle with gases neutralization with the help of calx and by removal of sediment with filtration.

Chapter 1
CHARACTERISTICS OF SHUNGITE ROCKS
AND GROUNDS FOR TECHNOLOGICAL WAYS
OF THEIR PROCESSING

1.1. Common Characteristics
of Zazhogino Deposit Shungite Rocks

Big group of diverse by their physico-chemical composition carbon-bearing pre-Cambrian rocks has got a title shungite rocks. This group
is unique in its mineral and chemical composition. The content of carbonaceous matter in these rocks ranges from 0.5 to 98 %, hence they can
be divided into rich and low-grade ones. Low-grade shungite rocks with
carbonaceous matter content from 0.5 to 5 % were called as shungite-containing, and with 5 to 20 % content — as low-shungitous [Kalinin, 1975;
Kalinin, 1984]. The rocks with carbonaceous matter from 20 to 55 % is
accepted to call shungite, their ratio in Zazhogino deposit is about 65 %;
the ratio of shungite rocks with C_{org} of 55—98 % reaches approximately
10—20 % [Borisov, 1956].

There are separate sections with carbonaceous matter content from
80 to 98 % in the deposit. The extent of these rocks is not large, they are
spread as a kind of small clusters along all deposit. While conducting min-
ing-geological works it is rather important to recede and store these rocks
separately as their applications spheres are rather specific. Rich in carbonaceous matter rocks are more homogeneous. Besides the carbonaceous
matter these rocks contain mainly fine dispersed earth silicon and just a little amount of other admixtures such as loam, aluminosilicate, carbonates
and iron-bearing minerals.

Shungite rocks with low content of carbonaceous matter from 1.0 to
15 % have inconstant material composition, hence it is important to classify
them with account of possible application area. Nevertheless, such classification by the processing technology of different kinds of shungite raw materials is absent. Big diversity of shungite rocks by the content of carbonaceous matter and other chemical components demands their classification
with the purpose to group them by branches of possible implementation.

For the first time shungite rocks were classified by Borisov P.A. [Bor-
isov, 1956]. This classification was simple and on the first stage of investigation, it turned out to be rather comfortable for practical application
(Table 1.1), it was used by the majority of scientists.

Table 1.1. **Shungite rocks classification** [Borisov, 1956]

Component, %	Shungite kind				
	I	II	III	IV	V
Carbonaceous matter	98	60	35	20	5–10
SiO$_2$	2	40	65	80	90–95

Following the classification, the shungite rocks were divided into five groups due to the content of carbonaceous matter and ingrained finely-dispersed earth silicon.

This classification shows that with the reduction of carbonaceous matter in the considered rock the content of earth silicon grows. In the frames of the deposit, mainly moderately-shungitous and shungitous rocks of three kinds (II, III, IV) are developed, their total extent constitutes approximately 70–80 %. All works to investigate shungite rocks were usually being conducted on following kinds: second. third and forth groups, while the attention was not being paid to rocks of first and fifth groups, and the area of their application has not been yet determined.

The major drawback of this classification was the absence of other chemical components within it, especially in shungite rocks with low-content carbonaceous matter. There is large quantity of bends, feldspars, carbonates, silicates, aluminosilicates, sulfides, which role in the industrial production is very big. These admixtures lend the rocks a series of new features that essentially influence the technological processes.

On the first stage of scientific investigations this classification had appeared to be major to develop all works on shungite rocks physico-chemical properties evaluation. At present, it has been collected the sufficiently large factual material on chemical composition and physical properties of all known low-grade as well as rich shungite rocks that are not included into this classification. That is why the necessity has been occurred to carry out principally novel classification, allowing to substantiate the structural and chemical properties of all possible shungite rocks. The novel technological classification of rich and low-grade shungite rocks with account of their chemical composition and technological properties will be considered in the Chapter 2.

Shungite rocks represent elementary carbon with globular and supramolecular structure that exits in metastable state, close to the form of technical carbon and white carbon. The shungite rocks differ from graphite ones by absence of crystal structure. They also differ from pitches and coals with low tenor of volatile matter components: oxygen, hydrogen, stink damp, Their approximate content in a whole in the deposit constitutes 1–2 %. The characteristic specificities of shungite rocks, their subtle concretion with other rock-making minerals, their qualitative- quantitative ratio define physico-chemical properties of shungite products, used in the industries.

The interest to shungite products as a new source of non-traditional raw is educed in following industrial branches: metallurgical, chemical, rubber, construction, vanish-and-paint and others. The experiments to apply shungite rocks in different industries were being carried out mainly on samples with carbonaceous matter content of about 25–40 % [Maslakov etc., 2005].

As it follows from various sources [Filippov, Romashkin, 1996], the application area of shungite products is very manifold and is not completely established; so far as industrial exploration of Zazhogino deposit it can be vastly widened.

The usage of shungite rocks that are low-grade by carbonaceous matter will be being defined after the mastering of industrial production under construction, taking into account of accumulated experience and statistical data, which will be being itemized during industrial introduction.

Zazhogino shungite rocks have been known for a long time, for more than 200 years. Initially, they were used by locals as paint [Kalinin, Gorlov 1968]. As it was mentioned above, the attempts to apply them as a power fuel with carbonaceous matter average content of 25–40 % did not meet the necessary efficiency because of high ash-content (about 70 %) of the rocks and bad capability to burn. Initially, all industrial experiments with shungite rocks were being pursued with the purpose to use them as a power fuel. These testings were fulfilled yet in 30's of last century, but were not brought to industrial introduction [Kryzhanovsky, 1931; Yagodkina, 1984].

There were attempts to apply the shungite rocks while the decoration of different industrial and residential buildings. The floors of Isaciy and Kazan cathedrals of Saint-Petersburg, some of the buildings in Petrozavodsk are decorated with shungite slabs. While the floor garnish of the Cathedral of the Redeemer in Moscow the shungite stone was used also.

Zahogino deposit is located on the West part of Onega Lake. The deposit stretches towards North-West direction on 20 km with the width of 4 km. The total square of deposit is 120 km^2 (Fig. 1.1).

The shungite rocks of the deposit are rich with carbonaceous matter and non-homogeneous in their structure and properties. Chemical composition of the rocks dramatically differ at different points of the deposit and also at different depths of their attitude.

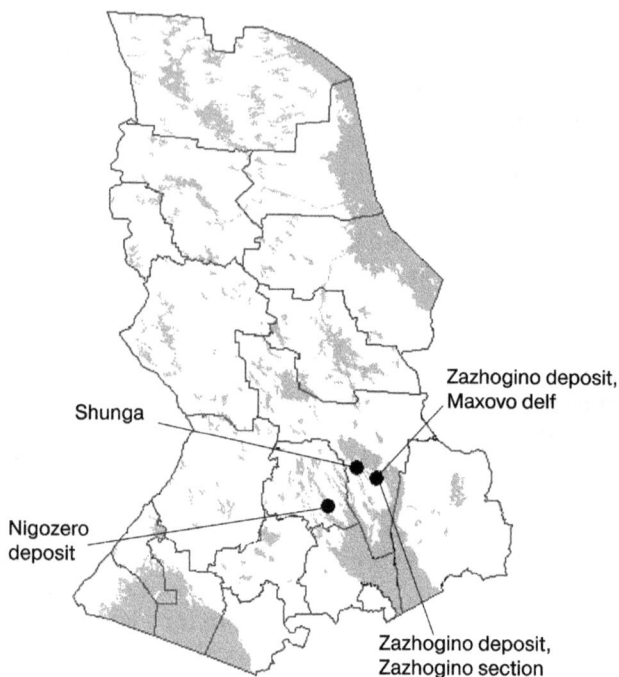

Shunga

Zazhogino deposit,
Maxovo delf

Nigozero
deposit

Zazhogino deposit,
Zazhogino section

Fig. 1.1. Location scheme of Zazhogino deposit

The deposit was explored in the details in 1989, and has been being exploited from 1974. Initially, the getter volume was 50 000 tons/year. Factually, the same volume was preserved till 2006, i.e. through almost 20 years.

The supply with electric power of all of the deposit is being carried out from high-tension electric transmission line of city Medvezhyegorsk (Karelia). The distance from main railway of city Medvezhyegorsk till the deposit is 70 km.

From all the available shungite deposits in Karelia (Table 1.2) Zazhogino is the most explored and prepared one for the industrial introduction. Its stated resources constitute 149,3 million tons, in particular, by categories: $B - 6,5$; $C_1 - 25,0$; $C_2 - 117,8$. The supposed yet non-stated resources is another 173 million tons. Besides, there are several promising sections in Medvezhyegorsk, Eastern Medvezhyegorsk, and Western. The supposed resource of these sections by geologists datum is about 1 billion tons. The mineral assets of these groups have been investigated yet scarcely, and their values are not stated.

Table 1.2. Explored Resources and Unexplored Assets of Shungite rocks of Onega Structure

		Deposits and sections									
		Weatern Kondopoga district				Zaonezhye Medvezhyegorsk district			EM & P*		
		Sandalsky section	Spassogubsky section	Kyappeselga section	Other spaces of shungite rocl evolution	Zazhogino deposit	Foymoguba deposit	Yandomozero section	Other spaces of shungite rocl evolution	Other spaces of shungite rocl evolution	Total
Resources, million tones	Total					149.3					149.3
	B					6.5					6.5
	C_1					25.0					25.0
	C_2					117.8					117.8
Assets, million tones	Total	88	50	50	80	173	200	200	150	50	1041
	P_1	4				124					128
	P_2	84				49	200	200			533
	P_3		50	50	80				150	50	380

* Eastern Medvezhyegorsk and districts.

At structural point the considered deposit represents compact fold (Fig. 1.2) with slope angle on wings from 10 to 70 °, mainly massive and brecciated rocks, here and there eroded and fissile. Into the common contour for exploitation, the shungite rocks with more than 20 % of carbonaceous matter are included. Their processing technology has not been worked out enough yet.

It is shown on Fig. 1.2b that the upper part of open-pit mine is deeply abraded and destroyed by weathering. In the upper part of the deposit space, massive rocks are developed, which total capacity constitutes 25–40 %. Massive rocks are also situated in the near-bottom of the deposit. Brecciated rocks are located in the middle and lower part of the delf, crossing line from massive to brecciated rocks is smooth. The volume of brecciated rocks is about 60 % of all the delf [Maslakov etc, 2005].

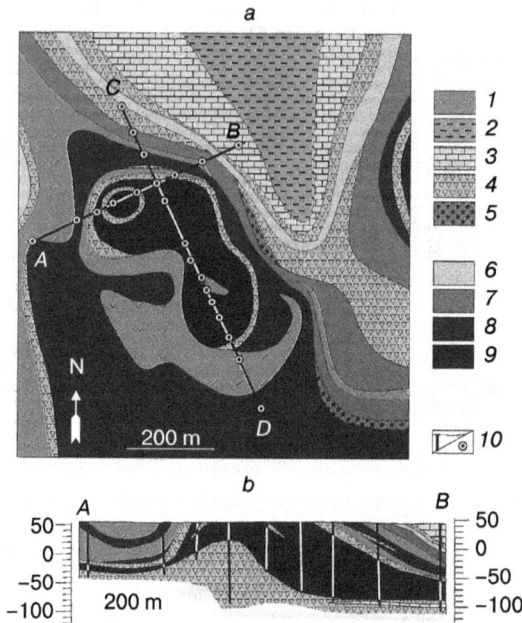

Fig. 1.2. Geological scheme (a) and open pit-mine (b) of Zazhogino deposit
(according to Kupryakov S.V. with some revisions):
1 — sills and dikes of gabbro-dolerites; *2* — silstones; *3* — carbon-
bearing rocks; *4* — basalt tuffs, tuff silstones; *5* — clints; *6–9* — levels
of shungite rocks; *10* — shafts

At the basal complex of the deposit, there are tuffs, tuff silstones, low-carbon and carbon-bearing shungite rocks. In the center of the fold under the deposit, there is quaquaversal body, consisting of lyddites and tuffs, which carbon-bearing tremolite metasomatites are well-developed on. Under all the deposit, delf, basalt tuffs, dikes and gabbro-dikes underlie.

The surface of all the deposit is broken-down. The grikes are filled with vein earth silicon, hydromica and brassil. Tiff, dolomite, feldspars and others encounter in the ettles. Anthraxolite always encounters in all of the available minerals there. The massive materials of the deposit upper part, i.e. uncovering, are very eroded, oxygenated and ochered. The height of uncovering rocks — from 0.2 to 8.0 m.

Taking into account the sub aerial occurrence of all shungite rock basins of Zazhogino deposit, the mining operations should be pursued only by surface mining, stripping down the thin layer of uncovering

12

ettles, consisting of earth silicon and also aluminosilicates and carbonates. There is small amount of carbonaceous matter (less than 5 %) in the uncovering rock. These rocks can be used as a raw for the production of low-grade shungite products.

At present, already for more than 20 years, an experimental area on shungite rock getter of Zazhogino deposit has been working. The increase in shungite rock getter can be easily managed by additional usage of 1–2 excavators and several pit-run dump tracks.

The running surface mining is the most cost-cutting, technologically simple and fully mastered at present. In the case of necessity, the getter volume can be easily extended with the account of projected capacities.

Besides the major delf of Zazhogino deposit, nearby, there are several low-grade, already explored sections of shungite rocks with carbonaceous matter content from 2 to 15 %. The occurrence of these sections is also surfacing that guarantees their getter by surface mining.

Additionally to main Zazhogino section, the experimental section "Mironovsky-1" is planned to be developed. At this section, almost all shungite rocks are low-grade, the content of carbonaceous matter there is from 1 to 15 %. Nevertheless, more prolific ettles with carbonaceous matter content from 30 to 40 % exist. The experimental works at "Mironovsky-1" section on shungite rock getter have not been started yet.

The license for this section mining by OJSC "MKK-Holding" (LLC "MKK-Engineering") has been obtained.

The brief analysis of shungite rocks of Zazhogino deposit has been showed that this deposit has been completely prepared for industrial exploitation. Sufficient number of different consumers for all kinds of production output exists.

1.2. Results Analysis of Shungite Rock Research.

Big complex of research on the study of composition and properties of shungite rocks, spheres of their application and also on the concentrating of low-grade shungite rocks has been pursued mostly by 3 research institutes: Mechanobr (years 1930–1933). Sciences Academy Institute of Geology Karelian Scientific Center (IG KarRC;years 1935–2005, administrated by Kalinin Y.K., High degree PhD in Technology, and Filippov M.M., High degree PhD in Geology-Mineralogy, Kovalevskiy V.V., High degree PhD in Geology-Mineralogy) and VIMS(years 1990–2006 administrated by Ozhogina E.G., High degree PhD in Geology-Mineralogy, Anufriyeva S.I, PhD in Chemistry, and Isayev V.I., Engineer).

The employees of these Institutes learned all major kinds of the ettles, their chemical mineral compositions as well as the structural properties were established. It has been elicited that the content of carbonaceous matter in all rock kinds alternate from 0.5 to 98 %. Such content scatter complicates much the technology of shungite rocks processing and demands their special classification by chemical content and properties with the aim to elucidate the new areas of application in various industrial branches.

It has been brought to light that in different deposit spots there are rocks with different chemical and mineral composition. This essentially simplifies the separate getter and processing of high- and low-grade carbonaceous matter rocks.

Taking into account that major mass of shungite rocks in the deposit center contains about 25–40 % of C_{org} and 45–70 % of quartz, on primary stage of the deposit development, VIMS adopted the decision to direct experimental-industrial production only for particular chemical content

Further, with the development of production basis, the additional study of all presented rock kinds, especially, low-grade shungite ones was pursued.

On the basis of these studies, the series of valuable features of shungite rocks has been defined. It has been done for the rocks, being used at the beginning of the deposit development yet in 80' s of previous century. While, because of the absence of industry production of shungite products their reduction to practice is proceeding slowly. Peculiar structure and physico-chemical properties of shungite rocks imply wide practical sphere of their usage. Let us analyze briefly these Institutes results of work.

For the first time, the comprehensive investigation of shungite rocks of Onega suite were conducted by Mechanobr yet in 1932. Their results are expounded in production reports and publication [Yagodkin, 1984], where the detailed description of mineral and chemical content of these rocks was given. It has been shown that across whole deposit their composition is inconstant, either carbonaceous matter or accompanying components vary, in particular, those of finely-dispersed silica, which present in different structural modifications.

The content of shungite rocks besides carbonaceous matter is following : carbonates, silicates, aluminosilicates, minerals of potassium, natrium, calcium and magnesium. The central part of the deposit is more homogeneous by rock content. The ratio of C_{org} is within 25–50 %, SiO_2 — 30–55 %. The average tenor of carbonaceous matter alternates from 0.5 to 98 %. Despite the significant scatter of the values, all kinds of shungite ettles have their sphere of application, Nevertheless, the tech-

14

nical specifications for final goods were not defined at just mentioned period of time. The rocks of Zazhogino deposit were divided into four groups: broken-up, massive, quartz-shungite breccias, including those with amygdules. Each rock group possessed their own properties.

It was shown with the help of X-Ray phase analysis that carbonaceous matter, being in shungite rock, existed mainly in amorphous state. Crystal carbon also presents but in negligible amount (1–2 %) mainly as a graphite.

The presence of big quantity of ash in shungite rocks (65–70 %) and small energetic value while their burn in regular fireplace did not give an opportunity to use these rocks as a technological fuel (the temperature of firing of shungite rock of Zazhogino deposit is from 450 to 600 °C). In the process of shungite rocks burning the flame was spreading slowly and faded quickly while just slight temperature fall. Hence, it was irrational to apply shungite rock as a fuel.

To reduce the ash-content of shungite rocks Mechanobr conducted a complex of jobs on their concentrating by gravitational methods yet in the early 40's of last century. The purpose of the jobs was in reduction of ash-content and increase of carbonaceous matter quality while burning. Nevertheless, as a result, there was a failure to obtain the concentrates with low ash-content. Likewise, decentish results had not been obtained while usage of flotation methods of concentration. The results of these investigations are described in the publication [Galkin, 1933].

Because of the lack of positive results the further realization of research on shungite rocks concentration was factually ceased. It should be mentioned that the workscope on shungite rocks investigation in pre-war (here "war" is the Great Patriotic War 1941–1945, the part of Second World War with the participation of the USSR) and after-war periods was minimal due to absence of finances for the deposit mastering. All the works at that time were directed to the study of carbonaceous matter as a power fuel, but it hadn't met the needs of industrial applications.

Nevertheless, the unique features of shungite rocks and the possibility to use them without concentration served a push to continue searching and geological survey works with the purpose to study chemical-mineralogy characteristics of these rocks and determination of their application area.

From the early 1960's, Academy of Sciences IG KarRC began to pursue vast works on the study of physico-chemical and geology-mineralogy characteristics of shungite rocks on the basis of specially created laboratory. Besides, administrated by USSR Ministry of Geology, the works to state the value of rocks in the deposit and the sphere of their applications began.

Table 1.3. Chemical Content of Shungite rocks
(according to IG KarRC, Russian Academy of Sciences (RAS))

Content, %	SiO_2	TiO_2	Al_2O_3	Fe_2O_3	FeO	MgO	CaO	Na_2O	K_2O	C_{org}	S_{com}
					Broken-up						
Average	47.04	0.25	4.16	1.13	0.42	0.57	0.08	0.117	1.225	44.57	0.38
Min	38.62	0.19	3.11	0.35	0.27	0.43	0.01	0.03	0.94	31.9	0.13
Max	61.02	0.38	5.6	2.06	0.52	0.88	0.14	0.52	1.56	53.32	0.93
					Massive						
Average	54.42	0.23	3.74	1.49	0.55	0.59	0.17	0.062	1.342	36.57	0.9
Min	41.82	0.14	2.1	0.37	0.14	0.21	0.07	0.03	0.58	22.17	0.11
Max	71.04	0.3	5.09	2.96	1.72	1.67	0.56	0.12	2.24	50.4	1.9
					Quartz-shungite breccias with amygdules						
Average	60.58	0.201	3.16	1.17	1.06	0.54	0.13	0.048	0.964	31.42	0.73
Min	44.57	0.14	2.27	0.21	0.45	0.26	0.01	0.02	0.6	22.8	0.2
Max	70.07	0.32	4	4.12	4.04	1.65	0.43	0.1	1.74	44.31	2.65
					Quartz-shungite breccias						
Average	62.44	0.182	2.96	1.07	0.42	0.44	0.09	0.039	0.789	31.04	0.38
Min	47.72	0.1	2.04	0.3	0.14	0.21	0.01	0.01	0.48	15.63	0.1
Max	76.8	0.26	4.14	3.14	0.87	0.87	0.29	0.08	1.51	46.84	1.32

As a result of these joint works the structure and properties of shungite rocks of Onega structure were investigated and the sphere of their usage was traced.

Chemical, mineral composition and features of shungite rocks were being studied for a long time in IG KarRC [Kalinin, 1975, 1984, 2002; Filippov, 2000]. On the basis of these works major geological-mineralogical parameters were established and preliminary spheres of the industrial introductions were brought to light [Filippov, 2000]. We consider necessary to give brief analysis of the obtained results and show the chemical (Table 1.3) and mineral composition (Table 1.4) of shungite rocks, which provided the basis for industry technological scheme. These analysis results became the basis for detailed research of the rocks and creation of primary base of consumers. The attempts to conduct additional concentration of the rocks have not given the positive results.

Broken-up rocks with carbonaceous matter content from 31.9 to 53.3 %, 44.57 % at average. They consist of quartz rocks with C_{org} from 38.62 % till 61.02 %, average — 47.04 %. Aluminum oxide content is from 3.11 to 5.6 %, average — 4.16 %. The quantity of other oxides — Fe_2O_3, FeO, MgO, CaO, Na_2O, K_2O, TiO_2, SO_3 — negligible and alternates from 0.2 to 1 %.

Table 1.4. Mineral Composition of Shungite Rocks
(according to IG KarRC, RAS)

Rock	Quartz	Micronite components (the calculation of mineral part), %			
		Σ pelitic minerals	Illite + mont-morillonite	Field spars	Carbonates
Shungites of first and second assises					
First assise	18.95	22.71	16.86	32.13	13.39
Second assise	66.82	19.49	17.17	3.49	4.81
Shungites with high content of carbonaceous matter					
Cherty	45.0	31.0	27.1	12.0	3.3
Sodium	19.5	18.2	13.4	48.3	3.3
Potassic	20.9	43.7	32.2	24.7	2.8
Shungites with relatively low content of carbonaceous matter					
Potassic	58.9	26.5	18.7	5.3	2.6
Sodium	21.5	25.8	14.3	41.6	2.0
Carbon-bearing	30.4	22.3	17.7	6.0	33.7

Massive rocks. The content of silicon oxide in these rocks increases and constitutes the values from 41.82 to 71.04 %, average — 54.42 %, and carbonaceous matter content with the comparison of broken-up rocks decreases from 44.57 to 36.57 %.

In brecciated rocks (quartz-shungite breccias with amygdules) the average SiO_2 content increases to 60.58 %, and carbonaceous matter one decreases to 31.42 %.

The most low-grade by carbonaceous matter content (31.04 %) ones are quartz-shungite breccias. The ratio of SiO_2 in these rocks rises to 62.44 %.

Thus, all shungite rock varieties are enough rich with carbonaceous matter and earth silicon. Almost in all prolific rocks, the number of mineral admixtures is slight, hence, their usage can be conducted without preliminary concentration. Low-grade rocks contain much quantity of admixtures, and while their usage they should be additionally enriched.

The mineral composition of shungitre rocks by deposit is presented in Table 1.4. As it is seen, besides the carbonaceous matter they have in %: earth silicon from 18.95 to 66.82, pelitic minerals 18.2–43.7, including illite+montmorillonite at everage of 13.4–32.2, field spars 3.49–48.3, carbonates 2–33.7.

The shungite rocks of main assises (first and second) in Zazhogino deposit drastically differ in the ratio of finely-dispersed earth silicon: 18.95 % — in the first main assise, 66.82 %. — in the second one. Such sounding alternations in silicon content can lead to dramatic change of physico-chemical properties of the rocks. Pelitic minerals, including

17

illite and montmorillonite, reside in assises approximately in the equal quantities: 22.71–16.86 % — in the first assise, 19.49–17.17 % — in the second one. Similar variations say about essential difference in physico-chemical properties of considered ettles. During the process of their get-ter and processing, the content of field spars in the rock should be strictly controlled, as they influence forthcoming technological processes much.

The content of carbonates of calcium and magnesium across the de-posit also vary: 13.39 % — in the first assise, 4.81 % — in the second one.

The outlined differences in mineral composition of the first and second assises surely will influence physico-chemical properties of the rocks that can lead to unstable results during their industrial application. For this rea-son, during these rocks getter, it is necessary to adjust their quality and also the possibility of getter selectiveness distinctly .

In the cherty ettles. the content of silicon oxide constitutes 35 %, in so-dium and potassium ettles — 19.5 and 20.9 %. Big number of pelitic miner-als or field spars present in these rocks. Thus, in sodium rocks, the content of field spars is 48.3 %, in potassium ones, — 24.7 %.

Carbonates in shungite rocks with large ratio of carbonaceous matter occur in small amounts and constitute approximately 3 %.

There is much earth silicon — 58.9 % in shungite rocks with relatively low carbonaceous matter content and high content of potassium. The con-tent of SiO_2 — 21.5 %, field spars — 41.6 %, in potassium ettles.

All of the above-considered rocks, we think, have different physico-chemical properties that will affect while their usage in different industrial branches. So, during technological processing, it should be necessary to collect statistical material on evaluation of all shungite rocks varieties. The material composition of shungite rocks with relatively low ratio of carbo-naceous matter — less than 10 % — is rather complex and peculiar. Large quantity of carbonates — less than 33.7 %, or aluminosilicates, while just 6 % of field spars can be in these rocks. The spheres of application of such rocks and rocks of the first and second assises will be dramatically different.

Low-carbon shungite rocks, as a rule, contain big amount of pelitic minerals — 40–45 %. The application area of such rocks will much vary from considered above. There are no yet consumers for these rocks, their industrial evaluation will be given after long testings in different conditions of their application.

As we see, the chemical and mineral composition of shungite rocks is rather manifold, and area of application of their varieties has not been con-clusively established. Therefore, special classification with account of their chemical and mineral composition is on demand. On the basis of the classifi-cation, their possible application area must be distinctly defined and adjusted.

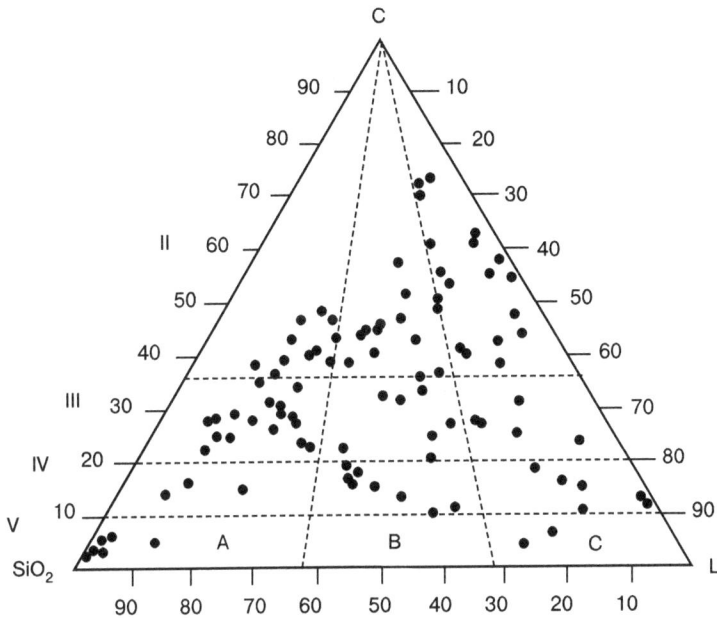

Fig. 1.3. Chemical- genetic classification of shungite rocks.
[Galdobina, Gordov, 1975]

To create such classification, all geological material, collected on the basis of previously made research works, conducted during last decades, should be analyzed.

For the substantiation of new classification the most interesting thing, from our point of view, has become the distribution of different shungite rock types as a triangle diagram (Fig. 1.3), worked out and justified in works [Galdobina, 1987; Galdobina, Gordov, 1975; Galdobin etc, 1986].

On this diagram, all kinds of rocks with different content of carbonaceous matter from group I to V (in Table 1.1) and with varying content of earth silica: A — high-, B — middle-, C — low-cherty. In high-cherty rocks (type A), SiO_2 content is more than 80 %; in middle-cherty rocks (type B) — 67–71 %, in group C it declines till 35 %. In the rocks of A and B types potassium silicates predominate, in the rocks of type C sodium as well as potassium constituents present. According to this diagram, we can outline that with maximal ratio of carbonaceous matter in the rock, free earth silica can be absent. Such rocks, as a rule, are referred to chemogenic ones and consist mainly of just carbonaceous matter with small admixture of earth silica that is contained in main matrix of carbonaceous matter. As far as we think, such rocks can be used even as a major power fuel and also as mineral fertilizers and natural mineral paints.

Due to described genetic classification, the accumulation of shungite matter proceeds as a result of its adsorption mainly in aluminosilicates.

All the shungite rocks, demonstrated on this triangle diagram, can influence differently consumer properties of shungite products. While, these questions have not been considered yet and demand substantial follow-on revision during industrial introduction of experimental-industrial plant in Lianozovo district (Moscow).

Last 15–20 years VIMS conducted large complex of scientific investigations on physico-chemical properties and structure of shungite rocks of Zazhogino deposit. The obtained results have utterly confirmed composition and features of shungite rocks that were established earlier by Mechanobr and IG KarRC, RAS.

VIMS geologists and technologists showed that Zazhogino and Maxovo deposits of shungite rocks consist mainly of carbonaceous matter and finely-dispersed earth silica of different structural modifications. As admixtures there are silicates, pelitic minerals, field spars, carbonates, aluminosilicates. As a rule, sulfide minerals present in all samples, collected at different points of the deposit. Their role during processing and industrial introduction is rather important. Average content of sulfide minerals across the deposit is from 2 to 3 %.

In the low-grade rocks, sulfide ratio usually alternates from 1 to 4 %, and in high-grade rocks, at some deposit sections, it can reach 5–6 %, and occasionally even 7 %

It is stated that finely-dispersed earth silica of the deposit is presented by its all possible varieties that nature has. These are fissile, chained, mixed and other crystals, which size varies from 0.1 to 50 microns. It points on that crystal structure of earth silica in the process of genetic metamorphoses was not fully formed. Therefore during thermal processing of shungite rock the crystal lattice transforms to more perfect one that proceeds in drastically reductive conditions. This effect has rather positive impact on the processes of dry metallurgy of iron stones and non-ferrous metals. For this reason, already first tests with shungite rocks showed good technological results in the processes of metallurgical fusion.

As mentioned above, besides the carbonaceous matter and finely-dispersed earth silica, carbonates in the form of tiff, brown spar, dolomite and also field spars, illites, iron sulfides feature shungite rocks. The presence of blinde, pentlandite, arsenic iron, chromic iron, apatite qua the secondary minerals is noted. Their dosage across the deposit is insignificant.

In the considered rock, the following micro-elements appear in small quantities: molybdenum, vanadium, nickel, chrome and copper. The average ratio of uranium is from 7 to $13 \cdot 10^{-4}$ %, and of thorium — $3.4–3.7 \cdot 10^{-4}$ %.

Such amount of uranium and thorium doesn't exceed safe limit norms, and shungite rock output can be used in all needed industry branches.

It was brought to light that in the rocks with higher sulfides content the quantity of micro-elements is maximal.

All shungite rocks were divided referring their structure into several groups, the same as in earlier research of VIMS was done: massive, brecciated and laminated (weathered, oxygenated). General ratio of brecciated rocks is 60 %, massive — approximately 40 %, laminated (weathered) rocks doesn't exceed 1 % [Maslakov etc., 2005].

Massive and laminated rocks are rather homogeneous relating to the content of major components ($C_{org} + SiO_2$). In more low-grade rocks, the ratio of admixtures — carbonates, aluminosilicates and others — grows several times up. The percentage of sulfide iron and sulfur in low-grade ettles is also variable in different parts of the deposit. There are much more carbonates and aluminosilicates in low-grade rocks.

Rocks, that are the richest in carbonaceous matter, contain essentially less silicates than low-grade rocks. The high-grade rocks also have less of iron and sulfides — these rocks can be used as a power fuel, mineral paints or mineral fertilizers.

The fissile rocks contain carbonaceous matter of approximately 20–60 %, and others are clays, aluminosilicates, carbonates.

The massive rocks as a rule have black color, high density and it is difficult to crush them with hummer. The massive rocks contain carbonaceous matter and finely-dispersed earth silica. The ratio of these components is approximately similar. The grain size of earth silica is from 0.04 to 0.15 mm.

It was established during the investigation of samples with an electronic microscope that earth silica grains form aggregates of different shapes. It was emphasized that separated runs occur in the shungite rocks. Inhomogeneously impregnated pyrite also presents in the massive rocks; the content of secondary earth silica constitutes less than 5 % of whole volume.

Thus, following conclusion can be done on the basis of research works analysis, pursued in above-mentioned institutes. The shungite rocks of all varieties posses the series of unique chemical and technological properties which can be widely used in different industry branches even without preliminary enrichment.

The application area of each ettle kind — massive (SM), brecciated (SB) and weathered (SW) — is necessary to establish before the beginning of industrial exploitation of plant under construction at present. The average chemical composition of all considered varieties according to VIMS data is shown in Table 1.5 and was taken into account in the project of half-industrial facility under construction at present.

Table 1.5. **Average chemical composition of different shungite rocks types**
(according to VIMS data)

Rock type	C_{org}	C_{com}	SiO_2	Al_2O_3	MgO	CaO	S_{com}	SO_4^{-2}	K_2O	Na_2O	Fe_2O_3
SM	31.43	31.62	51.8	4.54	0.73	0.06	1.33	0.14	1.30	0.31	3.54
SB	25.66	26.64	57.6	3.93	0.56	0.10	1.04	0.11	0.94	0.32	4.32
SW	63.02	67.44	10.07	3.40	0.01	0.03	1.18	–	0.31	2.47	1.63

Because the amount of massive and brecciated rocks constitutes almost 99 % from all Zazhogino deposit rocks as to VIMS data, an average chemical composition on the basis of these kinds has been adopted in the regulations.

The weathered rocks sharply differ by chemical composition, but because of small cubic capacity of these rocks (about 1 %) they are not being taken in account while the calculation of average chemical composition.

So, two kinds were taken as a basis of projected technology of shungite products industry. Both these kinds possess varying physico-chemical properties despite rather close chemical composition that must be accounted while their exploitation process.

The same technology of shungite rocks production has been accepted for all rock varieties on the plant being built. This is a great drawback of such technological scheme. For its more detailed assessment, let analyze this scheme and demonstrate its equipment arrangement.

1.3. Common Assessment of Physico-Chemical and Structural Specialities of Shungite Rocks

The general analysis of Zazhogino deposit shungite rocks has showed that, across all the deposit, they underlay rather compactly, but inhomogeneously by material composition. On certain spots, drastic deviation from the chemical composition accepted in the project is being observed. The impact of deviations in carbon and silicon dioxide contents in the shungite rocks on their consumer properties has not been investigated properly yet.

The major works on the study of influence of shungite products chemical composition have been carried out on the samples with average carbonaceous matter content from 25 to 40 %. In these rocks, big amount of finely-dispersed earth silica — less than 70 %. Such content across the deposit corresponds to technical regulations on the being included in the project end products. We can regard on that main shungite ettles of Zazhogino deposit can be utilized as a final output without preliminary refinement. Nevertheless, in each particular case, this question should be brushed up on with account of chemical properties of shungite products and spheres of their application.

The efficiency of usage of low-grade shungite rocks with carbonaceous matter content from 0.5 to 25 % and rich — from 40 to 95–98 % in different industry branches has not been enough studied yet, and the application spheres of such rocks have no production confirm. Accordingly, while the primary getter of shungite rocks, it is supposed to pursue the underground mining in a way to use those rocks where average content of carbonaceous matter in shungite products is within 25–40 %. In the future, while more detailed study of application area of shungite rocks, it will be necessary to investigate additionally the impact of chemical and mineral composition of particular varieties of the ettles on the quality of products being sold.

The average optimal chemical content of shungite rock, as to VIMS data, which considered processing technology was aimed on, is shown in Table 1.6.

Table 1.6. **Chemical compostion of shungite sroducts, sccepted in the project**

SiO_2	TiO_2	Al_2O_3	FeO	MgO	CaO	Na_2O	K_2O	S	C	H_2O_{cryst}
57.0	0.2	4.0	2.5	1.2	0.3	0.2	1.5	1.2	29.0	4.2

At present, the application area of shungite rocks in different industrial branches has been worked out on the ground of this chemical composition, basing on average statistics data that have been obtained in the process of laboratory and industrial experiments.

Consistent with adopted chemical characteristic the technical parameters of all shungite products have been carried out. These shungite products are planned to be obtained on being projected experimental-industrial facility. Main physico-chemical parameters of shungite products, put in the Project according to VIMS data, are shown in Table 1.7.

So, the obtaining of big number of shungite products is planned on the being constructed experimental-industrial enterprise. These shungite products differ between each other only by coarseness. Chemical composition is not taken into account in the transformation process. It is supposed that on the first stage of mastering of the experimental-industrial facility, only those shungite ettles will be delivered, that are in the center of main deposit, which chemical content will meet rock average chemical content, presented in the considered Project.

Table 1.7. **Technical characteristics of shungite rocks by the project**
(according to VIMS data)

Carbon content	25–40 %
Earth silica content	45–70 %
Silicate content	5–10 %
Firmness by Moos	4–5
Initial coarseness	less than 210 mm
Initial dampness	
average	6–8 %
maximal	11 %
Absolute density	1.8–2.1 kg/dm^3
Bulk mass of fragmenting	1.0–1.2 kg/dm^3
Bulk mass of powder	0.2–0.7 kg/dm^3
Natural angle of slope	
grit	30–45°
powder	20–30°

In the future, by accumulating the testing data, it is planned to conduct additional research works with the involvement of all kinds of low-grade rocks, which are in the deposit. These rocks are silicate, aluminosilicate, silicate-sodium, aluminosilicate-carbonate, carbonate and other possible ones, which can be found in the process of further investigations. All these rocks significantly differ by the chemical composition from the rocks included in the Project and, sure, their application area will significantly differ from the existing one. Additional conducting of large research work complex with establishment of possible sphere of their usage for these products is needed.

It is supposed that additional enrichment will be needed with account of new technical decisions, connected with firing usage, chemical leaching, autoclave processing. These works are not being pursued at present because of the limited financing. Hence, major direction of work are being considered — the multi-side usage of all shungite rock varieties, the high-grade as well as low-grade and the definition of major physico-chemical legitimacies of their possible application.

1.4. Aims and Tasks of Research on Advance in Present Technologies and Finding of New Technologies of Shungite Rocks Complex Usage

It was established by the first industrial tests on the usage of shungite products, finely-dispersed as well as coarse-grained, that the content of carbonaceous matter in the shungite ettle must be around 35–40 %, finely-dispersed earth silica — 45–70 %, alumino-silicates — 5–10 %, while the range of these components content across the deposit is much wider. For instance, the concentration of carbonaceous matter varies from 5 to 98 % and of earth silica from 10 to 90 %. The possibility to use industrially the shungite rocks with such wide range of chemical composition has not been stated, the mechanism of their interaction on different productions, including pyrometallurgical processes has not been studied. To use these rocks at industrial scale one should understand mechanism of redox reactions, which proceed in pyrometallurgy processes while the addition of shungite products with different content of major products — carbonaceous matter and earth silica. Also one should carry out the evaluation methodology of all the ettles varieties and establish possible spheres of their application in the industrial practice. For this purpose, the whole rocks volume should be divided into sorts. Rather full settlement of the problem would be possible only at long industrial usage period of shungite rocks with various content of carbonaceous matter and earth silica.

Some experiments in iron and non-ferrous metallurgy, having been conducted nowadays, are not sufficiently worked out despite the row of satisfying results. The activity mechanism of shungite rocks of different sorts in the process of their industrial usage has not been substantiated.

The absence of industrial plant for the production of shungite products is restraining the perspective of their practical usage, is complicating the study of their activity mechanism that is narrowing their application area.

At present the application area of shungite products with varying ratio of carbonaceous matter and finely-dispersed earth silica has not been completely determined yet. The physico-chemical impact of these rocks in the process of subsequent technological operations while industrial usage has not been sufficiently substantiated yet. The first shungite products, which industrial testing was conducted on, were obtained on experimental-industrial facilities with the use of samples with carbonaceous matter content of 25–40 % and finely-dispersed earth silica one of 50–60 %. These samples had not always reflected the specificities of mineral and chemical content of deposit in a whole.

Physico-mechanical parameters of shungite products in many respects depend on their chemical and mineral structure.

The technology of fragmenting and all-sliming must be proved and unified. It is also necessary to choose the most acceptable technology equipment and state optimal technology scheme of shungite products of varying coarseness.

Since shungite products, that are different be their coarseness, have rather different perspectives of their application according to their own physico-chemical properties, it is necessary to create industrial plants in near years on the production of shungite products that are stable by their chemical, grain-size and mineral composition.

Taking into account that reserves major volume of the majority of deposits is rather homogeneous by mineral content and physico-chemical features, VNIMS adopted the decision to apply Zazhogino deposit on the first industrial introduction stage without rocks additional enrichment. Hereafter, as far as the production development and involvement of other rock types of this deposit into processing, it is supposed to use additional enrichment operations, especially of low-grade shungite rocks.

The main task of the processing — is the creation of shungite products with different levels of coarseness and chemical content. Thus, for example, for metallurgy and constructing industries, the products with coarseness from1 to 10 mm or larger — from 10 to 100 mm are needed. When by way of filler in industrial-rubber production, the product with coarseness of 5–10, 15–20, 50–75 microns is needed. While at pigment production, the maximal particles size shouldn't exceed 10 microns.

As we see, the range of coarseness of used shungite products is rather wide, but their major application spheres are defined, first of all, by their chemical composition.

It has been determined with industrial practice that shungite products sometimes have better advantages by their physico-chemical properties in comparison with used today well-known products, for instance, for industrial-rubber goods such as technical carbon and carbon white.

Finely-dispersed shungite products have found wide application in various technological processes of chemical industry that will allow to expand the area of their application and to improve the technical-economical rates of working plants.

Nevertheless, despite the perspective of shungite rocks application, the scientifically grounded recommendations on technology choice for production of shungite products with different levels of coarseness and chemical composition are still absent. Because, as it was mentioned above, not all shungite rocks meet the industrial demands upon these indications.

Technological evaluation of all available varieties of shungite rocks and definition of novel spheres of their possible application must be based on their classification with account of chemical and mineral content. New technology of comprehensive development of all shungite rock kinds should be carried out on the classification basis. For this, the conducting of additional research works is at demand. Such works will allow to extend the practical application sphere of shungite rocks, particularly, those with low content ratio of carbonaceous matter. The settlement of the questions of complex usage of shungite products will allow to cut the self-cost of shungite goods production and drastically expand the sphere of their application.

The industrial advance of shungite products with broad range of variations in their chemical composition can lead to their irrational usage in different industrial branches. Therefore, primarily, permissible ranges of chemical composition alternations must be investigated and practically confirmed on the basis of industrial testing data.

Shungite products oxidative properties have very large significance while their usage. It has been noticed that during storage of shungite products, their oxidation takes place; therewith, the oxidation degree increases with the increase of the obtained products storage period. Hence, the task has appeared to figure out the formation mechanism of oxidating compounds, to settle their composition and properties and also to state the possibility to eliminate them from all shungite rocks varieties. The settlement of these questions will allow to extend the application area of shungite products in different industry branches significantly.

The major goal of present research is to study physico-chemical properties of shungite rocks for specification and expansion of their application spheres, and also for elaboration of production technology parameters and deletion of harmful admixtures such as acids, carbonates and others.

It should be noted that major specificities of application technologies of different chemical structure shungite products have been scantily learned. The activity mechanism of these technologies has not been established. The chemical composition of shungite rocks that should be considered in different application spheres has not been learned. The possibilities to use low-grade shungite rocks in various technological processes are not clear. The information on the ways of further development of processing technology and implication of low-sorted rocks into industrial turnover are lacking.

To clarify the efficiency of application process of shungite products in different industries and to establish the optimal areas of their application and also to find out the most reasonable work regimes of technological process the following task have been put forward:

1. To learn physico-chemical properties of all shungite rock kinds, also to observe the formation mechanism of acidic properties of these rocks and to establish their role in the manufacturing process.

2. To work out and theoretically ground the novel classification of low-
   and high-grade shungite rocks basing on their chemical content and
   properties. To give, on the classification basis, the complex assessment of all known kinds of these rocks.
3. To develop the mechanism and technology of elimination of harmful
   admixtures from shungite ettles by means of usage of chemical leaching processes.
4. To investigate the peculiarities of drying and all-sliming with substantiation of optimal technological parameters of these processes
   from the point of clean-up of midair.
5. To mastermind the novel directions of complex usage of low- and
   high-grade shungite ettles.

The current work is based on the scientific research results as well as
the industrial and laboratory experimental data. To study the composition and properties of the ettles the precise physics-chemical analysis
methods were used, including X-Ray, thermogravimetric, differential-
scanning calorimetry ones. The experimental data were then mathematically processed and modeled. The new methods were applied on determination of differential and integral distribution curves of differing kinds
of coarseness with the help of laser diffraction microanalyser "Analysette
Compact", Fritsch (Germany).

This allowed to get the new data, which will significantly widen the
present views on shungite products role in different industrial branches
and will allow to substantiate the possibility of their complex usage.

Conclusions

In a whole, observing the geological structure and chemical composition of Zazhogino deposit rocks, one can note the following particular
features appeared during its development and definition of application
technical parameters.

1. Zazhogino deposit is very promising for shungite ettles getter without
   preliminary refinement. The average concentration of carbonaceous
   matter in shungite rocks — 25–40 %, earth silica — 45–70 %. At
   bottom, this is ready concentrated shungite product, which is used
   in construction, metallurgy, chemical, rubber industries. The lower
   grade shungite rock kinds can be alloyed to rich ones that will also
   allow to increase ettle volume with rather high content of carbonaceous matter.
2. The getter of the deposit shungite rocks can be easily embodied by
   open-pit mining. Their compact bedding and low capacity of uncov-

28

ering (from 0.2 to 8.0 m) confirm that we can speak about rather low self-cost of these rock getter technology. Taking into account that at the deposit, during large period of time (1974–2007), the works on industrial getter of shungite rocks were being pursued throughout 3–5 tons/day with single shift employees, the further gain of ettle getter by 2–3 times up to the volume of 13–15 tones/day doesn't represent any difficulty and can be easily reached by means of additional delivery of winning excavator and other machines for rock ransporting.

3. The shungite rocks in the deposit center are rather homogeneous by chemical content; the concentration of carbonaceous matter alternates from 25 to 55 %. On spot sections of the deposit, there are observed significant alternations of the content of major chemical components: carbonaceous matter and earth silica (for example, C_{org} from 1 to 98 %). For the same sections, there are in evidence the drastic changes of other shugite rock components: silicates, pelitic minerals, field spars, carbonates and others. The ettles from these sections demand additional research with the aim of their possible development and application area.

4. The low-grade shungite rocks contain big amount of chemical admixtures, which content depends on the ettle location and properties. The impact of these admixtures on shungite products has not been enough defined yet. Their role will be being learned while the use of shungite products in the process of industrial introduction to various industry branches.

5. The shuhgute rocks that are rich with carbonaceous matter (more than 60 %) are more homogeneous by their chemical composition, their physico-chemical properties are constant. The volume of such ettles is inessential across the deposit and makes 10–15 %. In the industrial production, these rocks can be used as a power fuel, mineral fertilizer, organic paint and pharmaceuticals.

6. Generalizing the research results, obtained in industrial and laboratory environments, on production technology of shungite goods, and analyzing their application sphere in different industrial branches, VIMS together with LLC "MKK-Engineering" has carried out the regulations, and industrial designing and construction of the first experimental-industrial production (with dry technology) has been begun. The assembling of production major technological equipment finished in 2008. Start-and-adjustment jobs on the development of experimental production has begun.

7. The considered processing technology is built upon obtaining of shungite products of varying coarseness only. While their chemical composition is not taken into account despite its essential variations. Therefore, for the detailed evaluation of deposit all rock types, complex classification of all ettle varieties by chemical content and properties should be worked out.

8. The further works for shungite rock development must be all-up pursued not only with account of coarseness but also chemical content of all rock varieties. This matter will be discussed in details herein.

9. On the basis of all-up approach, the mechanism and technology of new products obtaining must be worked out that will allow to activate greatly the consumer market and improve technical-economical indexes of the plants.

Chapter 2
PHYSICO-CHEMICAL AND STRUCTURAL PE-
CULIARS OF SHUNGITE ROCKS
AND DIRECTIONS ELABORATION
OF THEIR COMPLEX USAGE

2.1. Common Vision on Chemical Properties
of Shungite Rocks

From the start of industrial application of various by coarseness shungite products, the series of their new, rather specific physico-chemical features has been figured out. Activity mechanism of these features and their role in different industrial branches have not been studied yet. While in the course of industrial absorption of shungite rocks in different branches, more and more new qualitative physico-chemical and technological characteristics occur that have not been known at all before and that should be experimentally proved and theoretically substantiated.

Carbonaceous matter, contained in shungite ettle, much differs from carbon of coal, graphite, artificial carbon and other analogous products which are widely used in the industry at present.

During the storage process of rich shungite as well as low-grade shungitous products, as a rule, the increase of their acidity that in the row of cases negatively impacts shungite products features. However. the formation mechanism of acidic properties of all shungite product varieties has not been learned yet, also, the reason of such acidity growth in correlation with storage time has not been clarified.

Thanks to practice [Kalinin, 2002; Maslakov etc., 2005; Filippov, Romashkin, 1996], it has been established that richer with carbonaceous matter shungite rocks have relatively more acidic response than low-grade and very rich shungite rocks. The shungite rocks with carbonaceous matter percentage from 25 to 65 % possess the highest acidity. The rich with carbonaceous matter products show high screening capacity during their usage as a wet plaster mix in construction. These ettles demonstrate the largest effect while their usage as mud bathes for to cure radiculitus, diseases of skin, musculoskeletal system, central nervous system and so on [Orlov, 2005; Rysiev, 2003. 2004]. As it is seen, physico-chemical properties of all shungite rock kinds cause not only satisfactory influence on human organism but also demonstrate themselves in the series of industrial branches: tyre, metallurgy, chemical, construction, electric, In these branches, already at present, industrial testings have been conducted that have given positive effect.

Accordingly, for some production branches, the work-out of optimization technology of these properties, particularly, acidity, and in some cases its complete elimination are in need. Meanwhile, acidity appearance mechanism has not been sufficiently studied yet; there are just several suppositions in literature, sometimes lowering the usage sphere of shungite products.

Presently, to neutralize the acidity LLC "MKK- Engineering" has carried out the technology for injection of special admixtures-modifiers into shungite goods that lower the acid formation process or eliminates it partially. These works have been worked out by Vilshansky A.I. As such modifying admixtures soda (Na_2CO_3), calcium carbonate $(CaCO_3)$, talc $(Mg[Si_4O_{10}](OH)_2)$ are used.

Usually during the fragmentation and fine powdering processes, the shungite products, as a rule, have signified acidic properties, aqueous extract pH ranges from 3 to 6. The increase of acidity is connected with excessive part of sulfur-containing components such as sodium and potassium sulfide, carbon sulphur, and also sulfides of iron and other metals. The negative influence of these components is felt particularly during the application of shungite products in industrial rubber goods, at which expense the rubber vulcanization time grows up. Thus, in the change of white carbon to shungite product the vulcanization time has grown. As it was emphasized to neutralize acidic properties leachate filler-modifiers were used, such as soda, calcium carbonate, talc; nevertheless such modifiers can cause negative influence on rubber quality. This matter has not been decided once for all yet and demands production confirm on industrial rubber goods plants.

All these admixtures, despite the obtaioned positive effect, do not ultimately solve and do not eliminate the acid formation problem for shungite rocks. Thus, during the check of modified shungite ettles on Moscow tyre plant, it was noticed the increase of vulcanization time from 5 to 6 % that is not desired thing in terms of technology.

To learn the activity mechanism of shungite matter in different production branches, first of all, it is necessary to learn chemical and mineral composition of all components that constitute all shungite-containing composite, and on the basis of obtained data, to work out physico-chemical bases of shungite products interaction in different industrial branches. As it was mentioned, rich shungite rocks consist of two major components — carbonaceous matter and finely-dispersed earth silica. As an admixture, one can find there silicates, aluminosilicates, carbonates, sulfur-bearing minerals, clays, field spars, micas and others. The ratio of accompanying minerals in the rich and low-grade rocks of different deposit sections is drastically varying, and, surely, ettles physicochemical properties change that influence the technological parameters while these ettles usage.

To study the physico-chemical properties of various kinds of shungite rocks it is necessary to consider all complex of their mineral constituents and by these means to define the possible area of their application in different production branches such as industrial rubber, metallurgy, chemical, construction and others.

Hereat below, the process of shungite rocks forming will be discussed and, on the obtained data basis, the possible impact of their features on technological indexes will be defined.

2.2. Shungite Rocks Formation Mechanism and Its Impact on Technological Processes

Basically, two components of shungite rocks — carbonaceous matter and earth silica — define their major properties and physico-chemical and exploitation specialties. The presence of earth silica in carbonaceous matter matrix and its structural forms, particularly, coarseness, influences technological processes. Earth silica role in technological processes is determined by group of factors, influencing its formation, such as pressure, temperature, recrystallization processes and others.

Practical interest to shungite rocks had sporadic character for a long time. The periods of big interest to shungite rocks application were changing to periods of drastic drop.

In the works [Kalinin, 2002; Kalinin, Kovalevsky, 1977; Maslakov and others, 2005] with the help of different physico-chemical methods, it was stated that carbonaceous matter of different shungite ettles varies from regular coals — it is more close to anthraxolite and albertite groups. Carbonaceous matter of shungite rocks is more similar to Doneczk coking coals but they are more ancient. These are organic products, created from lake sapropels that were changing under different environmental influences during long-term aggregation of more than 2 billion years.

In the process of geochemical passages, sapropel matter went through gellation stage, probably, it was exposed to molted lava impact and then to quick cooling that rocks cleavage and their unique structures as well as the presence of finely-dispersed secondary mineral enclaves, which till present time have been in the carbonaceous matter by way of thin veins of finely granular earth silica and other compositions of shungite rock, points on.

The coarseness of earth silica particles in carbonaceous matter matrix and its structural peculiarities greatly differ in some kinds of shungite and shungitous rocks. Besides sapropels, the shungite rock carbonaceous matter comprises of organic mineral constituents of bitumens, asphaltenes and oils which, as a result of volcanic activity, were joining already formed sapropels. The process of carbonaceous matter transformation into the shungite rock solid phase was passing through the stage of colloid half-liquid state. During the process of colloids transformation and further quick cooling, the carbonaceous matter becomes broken-up, and grikes are being filled then with silicates, carbonates, field spars, pelitic minerals and others [Kryzhanovsky, 1931a].

In the process of quick cooling, the joining of bitumens, oils and others with sapropels also proceeds that had resulted in that the carbonaceous matter of truly new structure had been formed from half-liquid colloidal phase, which matrix included carbon and finely-dispersed earth silica. This matter represents intricate system that much differs from all known carbon rocks. By chemical composition and properties, and also the genesis, the shungite rock carbonaceous matter the most closely relates to anthraxolite.

The carbonaceous constituent of all shungite rocks is approximately similar that is confirmed by chemical and physico-chemical analyses, which are being fulfilled for various samples. The major difference of all shungite rocks kinds is in mineral-admixtures composition and their ratio.

The carbonaceous matter and admixtures content in the rocks from different sections of the deposit differs critically; here and there, the presence of vanadium (0.8–1.4 %) is described, as well as of brass, nickel, molybdenum, selenium, cobalt, strontium. Various sulfides: pyrite, chalcosine, cuprite, cupropyrite, arsenic iron, antimony bloom, sodium and potassium sulfides- present in the carbonaceous matter.

Yet Kryzhanovsky [1931 a,b] was writing about the appearance of shungite rock carbonaceous matter.

The presence of rare elements in shungite rocks, for instance, of vanadium gives the possibility to check it with a view to use it as starting crude for production of metallic vanadium.

The modern and ancient sapropelites — are organic products of lake and sea occurrence, which are applied everywhere as one of the best fertilizers in the agriculture. The carbonaceous matter of shungite rocks is the product of deep processing of ancient sapropelitic drifts by means of redox processes in the course of metamorphism [Ryabov, 1933].

34

It has been stated by IG KarRC RAS and VIMS geologists with the help of various physico-chemical methods that the carbonaceous matter in shungite ettle has nothing in common with coals, it is closer to anthraxolite and albertite group. Probably, they are the sediment derivatives of sapropel types, being exposed to transformation during 1–2 billion years.

Besides the ancient sapropelites, the shungite rock carbonaceous matter, probably, comprises organic products from oils, resins, bitumens. As a result of interaction of these constituents over long period of time of metamorphic transformations, novel carbonaceous matter of shungite rocks was formed.

The processes of organic matter conversion can be studied with the help of modern precise physico-chemical research methods of all organic-mineral components of shungite rock as a well as ash obtained after the burn.

As it was said the grikes had appeared in shungite matter which were filled further with secondary minerals, possibly, of volcanogenic origin. Probably, carbon-bearing components of shungite rocks were sapropels and organic bitumen-like products of volcanic outbursts.

As it was settled by the geologists the shungite rocks are about 2 billion years old. On appearance, shungite rock carbonaceous matter reminds hard coal.

Several hypotheses of shungite matter formation process exist.

One researchers believe that big amount of organic remainders was accumulating in ancient shallow-water basin. In the process of their decaying and thickening, various phases of carbonaceous compounds were being created. It means that carbonaceous matter of shungite rocks has purely sediment nature.

Others [Bannikova, 1990; Bondar etc, 1987; Vassoyevich, 1973; Galdobina, 1987] believe that shungite rock carbon-bearing matter has volcanic origin, probably. Various chemical composition of shungite rocks is connected, presumingly, with different cooling velocities and involvement into organic matrix of new chemical elements and mineral compounds.

So far as technological aspect the preliminary enrichment of shungite rocks on the being built industrial plants is not provided for, the task of selective getter is appearing, possibly, with forthcoming averaging before the dispatch for the processing. Such operation is necessary for to obtain the shungite products with set chemical content. To solve this task the methods should be carried out to substantiate technological evaluation of shungite rock quality with account of structural and chemical changes. These methods will be described thereinafter.

The shungite ettle at high temperatures demonstrates active reducibility that allows to use it with high efficiency in iron and also nonferrous metallurgy.

The shungite rocks, together with diamonds, graphite and various coals create the group of regular carbon-bearing rocks. Also artificial carbonaceous products exist: chark, technical carbon, activated coal, technical carbon of various kinds.

All considered natural and artificial carbon-bearing products have their own sphere of application and can't be mutually substituted in industrial practice. Technical carbon is planned to be substituted for shungite products in industrial rubber goods. Also, the chark can be partially substituted for shungite products while the fusion of iron concentrates.

The carbonaceous constituent of shungite ettles has its own area of application and can't be substituted for other carbon-bearing products. The carbonaceous matter of shungite rocks- is the original carbonaceous matter form, possessing the row of valuable characteristics that are outlying for other carbon-bearing products. Almost all D.I. Mendeleev Table in low amounts is presented in shungite ettle. (Table 2.1).

It is known that rare and trace elements cause the influence on the live processes of animals and human. Specialists suppose that the shungite stone regulates redox processes in human organism and acts as an enzyme and even a catalyst thanks the microelements abundance in it [Doronina, 2004].

Table 2.1. **Average Content of Microelements in Shungite Rocks ($n \cdot 10^{-3}$ %)**
(According to VIMS Chemical Laboratory)

Cu	Zn	Ag	Ni	V	Co	Mo	As	Y	Yb	Cd	Sb	Hg	Cr	W	Sc	Zr	La	Pb	Ba
15	40	0.08	20	4	0.8	5	20	20	2	0.5	3	2	20	0.5	4	8	7	2	150

Last years, rare enzymes were observed in the shungite rock carbonaceous matter. They were called as fullerenes. Fullerenes role in the industrial practice has not been disclosed yet. Some specialists outline their large value while the effect on live organisms. Nevertheless, it has not been confirmed experimentally yet. The activity effect on human organism of the carbonaceous matter, mixed with water, was observed far ago [Orlov, 2004; Rysiev, 2003, 2004].

It has been shown that while mixing the shungite rock with the water the solution is forming which demonstrates the healing effect. The positive effect of this solution on human organism was noticed more than 200 years ago while its activity mechanism has not been studied enough. It has been shown that finely-dispersed shungite infusion in the water kills E. coli, neutralizes the effect of heavy metals and decomposes chlororganic compounds as well as ammonia, nitrates and nitrites [Orlov, 2004;

Rysiev, 2003; 2004]. The neutralization process of these compounds is connected with large absorption capacity of shungite rock organic matter, therefore they have been started to use as an adsorbent during the purification of waste water from heavy metals, organic compounds and other harmful substances. Also, big amount of household filters has appeared where the finely-dispersed shungite rock is used as a major adsorbent.

It was not accidental that on the territory of shungite rock deposit, the first health resort in Russia is located. The resort was opened yet be Peter the I in 1719 — "Marcialniye vodi" where the musculoskeletal, urogenital and cardiovascular diseases are being healed.

Thus, the shungite rock carbonaceous matter purifies waters nicely as well as probably contains substances which allow to improve human health. The mechanism of healthy effect of shungite rocks has not been established yet. As we see, the shungite rock carbonaceous matter possesses the unique properties, making them different from hard coal, graphite, bitumen, oils and other carbon-bearing products. The carbonaceous matter in shungite rock acts originally. Apparently, it is connected with its unusual structure and also specific association of chemical elements.

X-Ray amorphous matter in shungite rock is presented by globular cumuli with 200–400 Å size of the globules.

It is assumed that namely globules are capable of separation from major mass of chippy shungite ettle while its shaking in the water. This property has been observed only in shungite rock. The number of globules in the water depends on the rock chopping degree — with the raise of chopping degree their number grows. Such shungite rock feature manifests essentially quicker on freshly bared surface. On the surface of durably stale chopped shungite rocks, the globules separation goes much slower and impact effect lowers. and water purification speed decreases.

From our point of view, the globules do not transform into the water but the regular process of leaching (oxidation) of sulfur-containing shungite rocks, mainly sulfides, proceeds there. Formed sulfur-acidic compounds by way of SO_2, SO_3 cause essential impact on further chemical processes.

Generalizing aforementioned one should notice that carbonaceous matter of shungite rocks and other carbon-bearing products, for instance, pit coal, natural gas, greatly differ by their physico-chemical properties. Shungite rock carbonaceous matter, as a rule, quicker enters into reaction in redox processes of fusion. As a result of carbonaceous matter oxidation the energy is releasing which is needed for life of plant and live organisms. The complicated evolution process in the plant world and live organism is unique and has not been studied yet to the end.

Almost all organic compounds contain carbon, and the scheme of their interaction and organic compounds structure are very manifold. Thus, for example, graphite has laminate structure, at that each layer consists of several modifications which contain triple bonds. Inside the layers, the bonds are more short, and they are essentially longer between the layers.

The graphite crystal lattice differs from diamond crystal structure, which represents correct crystal structure with even bonds. The distance between neighboring atoms in the diamond crystal lattice is minimal. The structure of lattice is compact, thereby diamond possesses high strength and firmness.

"Carbon", artificial carbonaceous material, contains double and triple bonds and therefore drastically differ from diamond. Hence, synthesizing the "carbon" by special technology one can perform a diamond.

In 1973, the domestic chemists Bochvar D.A. and Galperin E.G. figured out that carbonaceous matter of globular structure consists of 60 carbon atoms. And even earlier, abroad, in the journal "New Scientifist", the work of D. Jones was published where he showed that molecule of carbon consists of few graphite layers. It was figured out that spherical carbon shell consists of penta- and hexa-gons. By its structure, this spherical shell reminds the tyre of soccer ball or spherical roof. Similar roofs were being projected by American architect Richard Buckminster Fuller, therefore spherical structure of carbonaceous matter was called fullerene by alien scientists.

Such form soon was managed to be synthesized from graphite by means of its thermal decomposition. On the basis of these investigations the new name of carbonaceous matter had appeared — fullerenes. They have high stability and hollow ball-like (globular) structure which molecules contain 60 carbon atoms. The structure of globular carbonaceous matter of shungite rock is presented on Fig. 2.1.

In a whole the structural surface of carbonaceous matter molecule — fullerene- contains 12 regular pentagons and 20 and 20 scalenous sexagons. In the presented globular structure 90 sides take place which are in truncated icosahedron. These sides have same length. The carbon molecule of globular particle of shungite rock has been mathematically calculated and has been registered as a cluster with molecular number 60 (C60). The formula calculation and its substantiation were pursued by American scientists Harold. W. Kroto, Robert F. Curl and Richard E. The structure obtained mathematically has spherical surface with hollow interior space. In many countries there was a buss on obtained carbon molecular shape because it has allowed to explain all the gathered practical material on carbon-bearing products of shungite ettles.

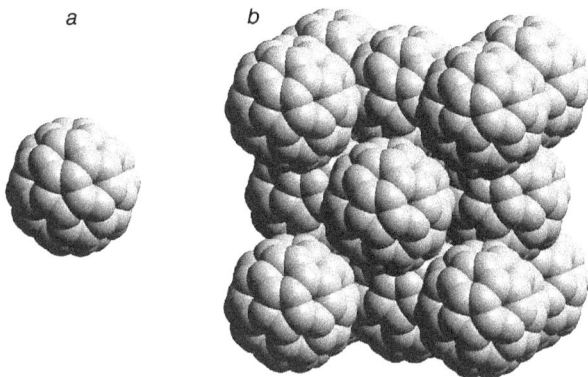

Fig. 2.1. Globular structure of carbonaceous matter molecule C_{60}:
a — fullerene; *b* — fullirite

The structure of carbonaceous matter, consisting of hexagons in shungite is analogous to crystal structure of graphite. Each carbon atom simultaneously belongs to 2 hexagons and one pentagon. The obtained data on carbon molecule composition served a push to develop a science about carbon, as these data had allowed to substantiate new forms of existence of carbonaceous compounds and their specific properties.

In the spherical carbon molecule shell, fullerene, as recent research of different countries scientists shows, big number of various chemical elements exist that leads to diversity of synthesized compounds properties. This is the new direction of organic and metalorganic compounds synthesis. On the basis of these works, several types of lasers of superconducting materials and microelectronic devices were created as well as the possibility to widen the application sphere of shungite rocks in new various production branches has appeared.

The seconds by its importance and quantity the chemical compound of shungite rock, defining its physico-chemical features, is the finely-dispersed earth silicon, which presents in the carbonaceous matter in different modifications.

The content of finely dispersed earth silica and other siliceous compounds keenly changes, especially, in the low-grade shungite rocks. Common content of silica-bearing minerals at different sections of the deposit much alternates — from 1.0 till 70–80 %, and, surely, chemical and physico-chemical properties of shungite rocks, delivered for processing from various sections of the deposit, differ. Therefore it is necessary to state the role of finely-dispersed earth silica and other silica constituents of shungite rocks in the technological processes.

39

At present, the mechanism of finely-dispersed earth silica has been insufficiently disclosed, in particular, its impact on reductive processes during pyrometallurgical ironstone processing. Also, the influence of silicates content on the kinetics of complex processes in pyrometallurgy has not been explained. These questions demand experimental investigations, including industrial ones, while the use of rich or low-grade shungite rocks in various industrial processes.

The earth silica and other silicate minerals, as the industrial tests have shown, possess high redox capability, especially while the use of shungite products instead of charred coal and silicon-containing additives in conditions of high-temperature metallurgical fusion.

Probably, at temperatures higher than 1000 °C in the pyrometallurgy processes, the recrystallization of finely-dispersed earth silica, contained in carbonaceous matter matrix, takes place. Usually, various modifications of silica dioxide, including metastable ones, present in the shungite rocks.

In the pyrometallurgy process, the crystal lattice of silica particles with defective structure rebuilds with forming of steady, stable structures. In the work of [Micyuk, 1980], it is shown that crystal structure lattice of finely dispersed earth silica, that is in shungite rock, depends, in primis on magma temperature and the speed of its cooling. Also the composition of magma molten influence the crystallization process.

As it has been noted already, the finely dispersed earth silica in shungite rocks presents in different polymorphic modifications that, undoubtedly, affects the process of their pyrometallurgy reprocessing. Consequently, let us briefly dwell on some specificities of siliceous mineral structure before considering the activity mechanism of finely-dispersed earth silica in technological redox processes.

As it is known, siliceous minerals are presented by whole series of mineral kinds (earth silica, christobalite, tridymite and others), stable at definite temperature and pressure intervals.

The 3-D skeleton makes the basis for their crystal structure; the skeleton is built with tetrahedron (SiO_4), connected via their mutually common oxygen atoms. Nevertheless, the symmetry of their location, packing density and mutual orientation are different that reflects the minerals physical properties. The siliceous minerals usually occur in the type of very tiny grains, often cryptocrystalline fibrous and spheroidical units, rarer in the type of tiny crystals of various shape. The most famous and widely spread siliceous mineral is earth silica.

Besides the siliceous minerals, the composition of shungite rocks includes various silicates — natural chemical compounds with complex silica-oxygen radical $[SiO_4]$. Their important peculiarity is in the capability

40

of polymerization that is very critical while the usage of shungite rocks in different technological processes.

By the character of silica-oxygen tetrahedrons binding one can distinguish island, ring, chain, laminate and skeleton silicates. All they are present in this or that amounts as a part of shungite rocks.

Island ones, or orthosilicates, are minerals of olivine, zircon, garnet and others; ring ones are beryls, cordierites and others; chain ones are widely spread in nature amphiboles, pyroxenes (it is noted the presence of enstatite — and also tremolite $Ca_2Mg_5Si_8O_{12}(OH)_2$).

The huge mineral admixture of shungite rocks relates to laminate silicates — micas, hydrous micas and other pelitic minerals, and also skeleton silicates, mainly, field spars, which maximal quantity is in the shungite rocks with Corg from 1 to 15 %.

The big changes in the structure of various minerals, presenting in shungite rock, are being caused by weathering process. Herewith, the minerals are being crushed, particle coarseness is decreasing, the volume and deepness of grikes is growing up, easily soluble compounds are being leached, the relative content of carbonaceous matter is increasing.

However, the amount of much weathered shungite rocks is not big and, according to VIMS data, is about 1 % from total rock volume.

Besides the major components of the shungite rocks, the carbonaceous matter and earth silica, iron-bearing minerals are widely developed there. The biggest iron content is referred to massive rocks where the Fe_2O_3 concentration is more than 3 %. The total content of iron oxide in other kinds of shungite rocks (laminate and brecciated) is no more than 2–2.5 %.

Iron-bearing minerals of shungite rocks, their chemical composition and structure are studied rather well. According to research results, accomplished by VIMS, they are presented by magnetite, hydrogoethite, pyrite (iron pyrite), bivalent iron silicates. The average percentage content of these minerals in shungite rocks of Zazhogino deposit is demonstrated in Table 2.2.

Table 2.2. **Content of iron-bearing minerals in shungite rocks**
(according to VIMS data)

Minerals	Structural varities of shungite rocks		
	Brecciated	Massive	Laminate-weathered
Magnetite	22	18	34
Hydrogoethite	26	19	23
Pyrite	40	55	41
Iron silicates Fe^{2+}	12	8	2

As it follows from the table, the maximal pyrite content is characteristic for massive shungite rocks (55 %); in these rocks other kinds, it is significantly less (about 40 %). The hydrogoethite content in different varieties of shungite rocks changes from 19 to 26 %. Approximately 65 % of all iron of shungite rocks is in sulfides and oxyhydroxides compositions which under the action of air oxygen, water and also temperature effects during the rocks drying are easily being oxygenized and degrading. The oxygenation and thermal degradation process of these minerals will be discussed below.

Other kinds of iron-containing minerals such as magnetite, silicates possess small chemical activity and almost do not degrade in the oxygenation processes under temperature, air and water action; their ratio in shungite matter is about 35 %. Magnetite fraction can be isolated from shungite rock with the help of dry magnetite enrichment, however, magnetite grains are very tiny and very fine pounding is necessary for their separation that is not always feasible because of low per cent content of iron.

Thus, it is impossible to eliminate the iron sulfides and oxyhydroxides with dry method from carbonaceous matrix of shungite rock; their role in the further usage of shungite products is pretty essential as, by means of these minerals, such product occurs which is swiftly oxygenated on the air while product pH lowers greatly from 6 to 3.5. In Zazhogino deposit, besides pyrite and mineral pyrite, there are arsenic pyrite, copper pyrite, pyrrhotine, molybdenite, cuprite and cinnabar. The presence of these minerals causes the influence on bacteriological state of the water, infused on shungite rock, which the row of well-spread microorganisms die in.

It's worth to mention that sulfides in shungite rocks of Zazhogino deposit are spread rather unevenly, they often fill the grikes in. The pyrite grain size varies from 0.074 to 0.005 mm, such particles constitute 85 % of all pyrite [Maslakov etc, 2005].

It is possible also to eliminate this fraction but only with wet gravitational enrichment methods, However, low percentage of its content (1—2 %) doesn't allow to use this method in industrial practice.

It is known that as a result of sulfide oxygenation, sulfuric acid is formed that has very big importance in the processing and consequent usage of shungite products. It is because, in some cases, the application of shungute products in industrial rubber production lows down the quality of obtained industrial rubber goods and also increases vulcanization time period.

2.3. Research of Physico-Chemical Properties of Shungite Rocks with the Help of Thermogravimetric Method

To study the thermal features of Zazhogino deposit shungite rocks thermo-analyzer (the device of synchronous thermal analysis STA-409 Netzsch, Germany) was used.

Mass

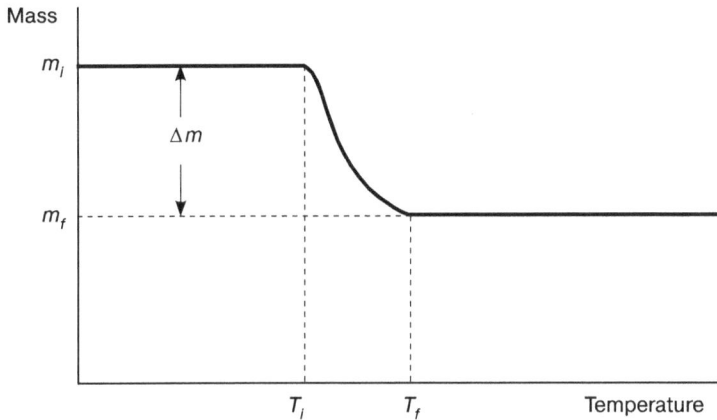

Fig. 2.2. The scheme of the matter mass loss curve.

With this thermos-analyzer, the synchronous changes of mass can be determined and possible calorimetric researches in the conditions of heating speed change at wide temperature interval in different gas medias. The researches were conducted in 20 to 1000 °C temperature range with speed of 10 °/min in firepots made of aluminum oxide without fresh air inflow. As a reference for comparing trials the aluminum oxide (ΔAl_2O_3) was used. The mass of initial weighment was 50 mg.

On the set device, the thermogravimetric influence can be studied — the method based on substance mass change depending on the temperature and time. Mass decrease degree depends on temperature and is recorded on diagram. The work principle of the device is shown on the typical curve of sample mass loss during single-stage decomposition of the matter (Fig. 2.2.). During the trial, the sample of mass m is being heated up with constant speed to temperature of T_i. whereby the decomposition of the sample is beginning and its mass is changing. Consequently, in the temperature T_i-T_f growth process, the further matter decomposition and its mass changing is proceeding. The values m_i, m_f and Δm are used for quantitative analysis of system body change.

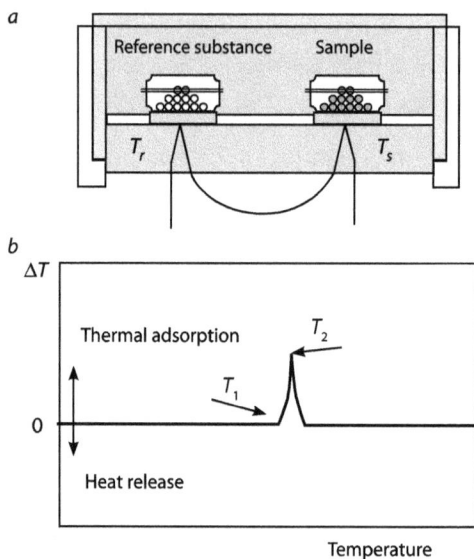

Fig. 2.3. The conductance scheme of differential thermal analysis (*a*)
and corresponding experimental curve (*b*)

With the help of differential-thermal analysis (DTA), the comparison of the sample and reference temperatures in the temperature preset range proceeds. The temperature of phase passages of the tested sample either remains short of the reference temperature — endothermic process, or outstrips it — exothermic process. Generally, the aluminum oxide or silica carbide is used as a reference.

The studied sample and reference are placed in one and the same block and are heated with preset speed. In each container, similar thermo-piles are placed which are connected between each other (Fig. 2,3a).

In the case if the tested sample and reference have the same temperatures, EMF of the thermopile also should be equal. In the case if, in the tested sample, there are the processes accompanied by the appearance of new heat effects, some difference in temperatures ΔT between tested and reference occurs that leads to the creation of some EMF amount on differential thermopile and to the release of additional heat. (Fig. 2.3b).

During the analysis the obtained square under each pick A is proportional to reaction product mass m and reaction enthalpy ΔH:

$$A = k \cdot m \cdot \Delta H,$$

where k is calibrating coefficient.

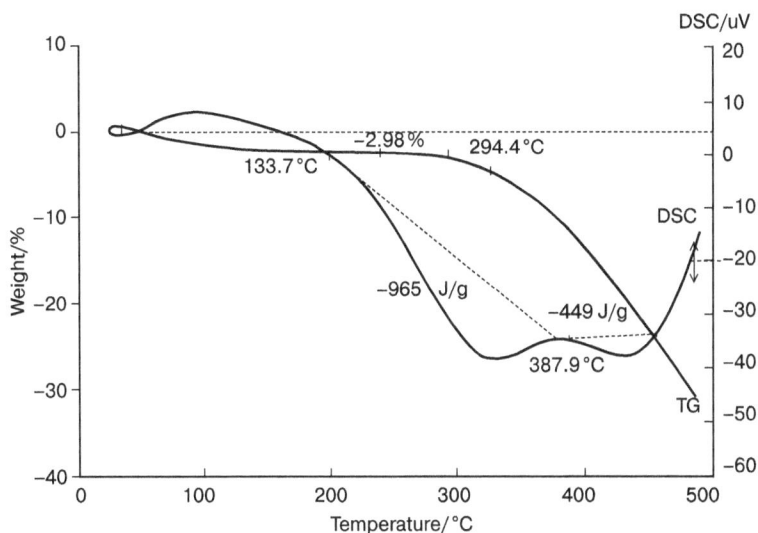

Fig. 2.4. Curves of differential-thermal analysis of rich laminate shungite rock

According to the equitation one can define the mass m in the case if other values are known. Being aware of them, one can calculate the reaction enthalpy from DTA and DSC — the calorimetry method which is based on the change of energy, not temperature.

The obtained DSC represents dependence of differential temperature $\Delta H/\Delta T$.

While the research, the definition technology was pursued in a way that the sample and the reference would get the same heat amount, i.e. the isothermal process would proceed. The measurements can be conducted with the help of two thermoelectric heating elements, which are coupled in one calorimeter. In the case if temperatures change, the heating speed also changes, and the temperature straightens out.

With the help of DSC method the substance content and its thermal characteristics are determined. TG and DSC curves for rich laminate shungite rock with carbonaceous matter content of 93.5% are presented on Fig. 2.4. Besides the carbonaceous matter, this product has small amount of silica oxide — 2.5% and about 3.5% of other constituents by way of sulfur and other gases H, N, O, CO. It has been figured out that in the heating process, thermogravimetric curve already at temperature of 50 °C, firstly, begins uniformly descent till 133.7 °C. Then, thermogravimetric curve goes parallel to x-axis with weight loss of 2.98 %, and next again gradually lessens till 330—350 °C temperature; while at further temperature growth, it continues to lower gradually.

At the beginning of heating process the sample weight descent is connected with the deletion from the rock of the gas constituents — nitrogen, oxygen. carbon oxide and sulfur. The elimination of these components proceeds evenly up-to temperature of 133.7 °C. While the 100 °C temperature has been reached the elimination of regular and crystallization dampness proceeds; with the temperature growth till 294.4 °C no any changes occur in shungute rock, and after further temperature increase the decomposition (oxygenation) of carbonaceous matter proceeds with CO_2 excretes.

DSC curve drastically differs from TG curve (in Fig. 2.4). From this curve, one can see that, with temperature increase, DSC curve is growing up to 100 °C temperature and then is starting to descent gradually till 200 °C.

With the temperature increase till 320 °C, curve is keenly falling down, and with temperature more than 320 °C is beginning to grow till 387.9 °C and then again is falling till 450 °C. During further temperature growth DSC curve is keenly increasing till 450—500 °C.

The initial DSC curve growth from 60 to 130 °C is connected with the decomposition of mineral crystal lattice which contains the crystal water — this is an endothermic effect connected with elimination of gas constituent of shungite rocks. Next temperature increase from 130 to 300 °C is connected with start of carbonaceous matter degradation and gas — carbon oxide — excretion. Next growth of DSC curve is connected with pyrite degradation and formation of iron oxide and sulfur vapor.

TG and DSC curves of other samples look like in some other way, the samples chemical composition is close to projected (Fig. 2.5 a, b, c). Thus, the sample 2 (Fig. 2.5a) — massive rock with carbonaceous matter content of 31.4 %, sample № 3 (Fig. 2.5b) — brecciated — 30.9 % and sample № 4 (Fig. 2.5c) — massive, held in water — 31.1 %. The behavior of these curves is approximately similar. TG curves of massive rock (sample № 2) and brecciated rock (sample № 3) are almost the same. TG curve of massive rocks at 180—200 °C temperature is falling, and then in temperature interval of 200—250 °C is beginning to grow. Such curve shape, presumably, is connected with sulfide decomposition and beginning of iron oxide formation because the pyrite content in sample № 2 was 2.9 %.

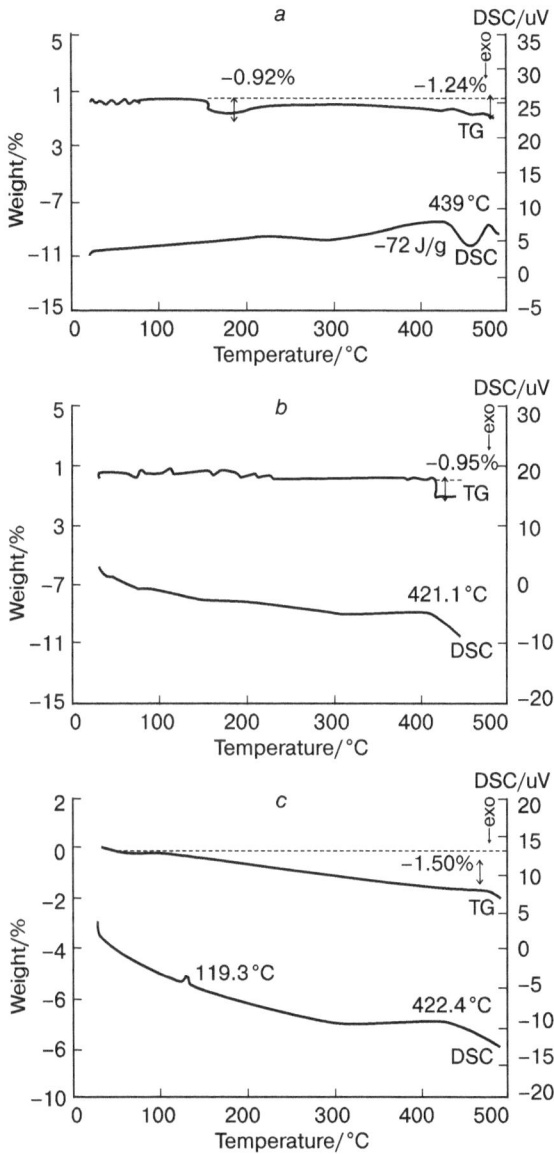

Fig. 2.5. Curves of differential-thermal analysis of:
a — massive shungite ettle; b — brecciated shungite ettle;
c — massive shungite ettle, held in the water

Then, with temperature growth till more than 250 °C, TG curve is beginning to fall slowly that also points on the beginning of carbonaceous matter denaturation and carbon dioxide formation.

TG curve in the sample № 3 of brecciated shungite rock looks in an other way. Here, this curve is almost parallel to x-axis, its small drastic fall takes place only at 400 °C temperature . This fall, apparently, points out the decomposition of other sulfides besides pyrite.

In the sample № 4 (the rock, has been held in the water) TG curve has started to fall evenly when reaching 40—50 °C temperature that is connected with the existence there, besides carbonaceous matter, of earth silica, which reacts on temperature growth while the absence of acidic constituents. Such curve behavior is connected with gradual deletion of regular crystallization dampness and even burning-out of carbonaceous matter.

In the sample № 4 in the process of shungite rock washing-off, almost all constituents, which create acidic reaction, were deleted. The carbonaceous matter, remained in common shungite rock mass, began to decompose gradually that has leaded to sample total weight fall.

DSC curve nature is essentially different in all three samples. In the massive shungite rock, DSC curve grows slowly, whilst it falls in brecciated and massive, held in the water rock (samples № 3 and № 4). This is a witnessing of that massive rock differs from brecciated one not only by structural composition but also by physico-chemical properties despite these probes chemical content is almost comparable.

The keen change of DSC curves character has been noted during the investigation of all three samples (№ 2, № 3, № 4) at temperatures about 420—480 °C. This points on that, at this temperature, other sulfide varieties (besides pyrite), available in shungite rock, begin to degrade,

Thus, learning the nature of heating curves in differential-thermal analysis, one can observe that according to self-structure and properties all the shungite rock varieties — massive, brecciated and laminate — have drastic physico-chemical differences and, surely, will cause original influence in the process of their industrial application. This influence should be taken into account and studied in each particular case individually with account of accumulated statistics experience. Thereupon, in the process of shungite rock industrial introduction, not only their chemical compositions but also their structural properties of all their varieties (massive. brecciated, laminate and others) should be taken into account.

2.4. The study of shungite rock structural properties by X-Ray method

The study of shungite rock structure was held by X-Ray method with the help of diffractometer DRON-3. On this device, the X-Ray tube of 0,7БСВ-2-Cr type was equipped. The digital and literal expressions of the tube were decoded by following way: 0.7 — is the tube capacity, Б — the tube in protecting cover with protection from X-Ray beams and electrically safe; С — tube for structural analysis; В — water cooling; 2 — windows number, Cr — the tube with chrome anode.

The samples of shungite fine-grained powder (1−15 microns) of 0.2−0.5 g weight were investigated. The intensity of X-Ray beams, being received by register, was measured by the way of radiance pulses reading:

$$n = N/T,$$

where N — is number of pulses, registered during time T.

The experiments were held on shungite rock 3 samples of following varieties: laminate, massive and brecciated.

Shungite rock laminate sample was taken intentionally with very large carbonaceous matter amount — 95.8 %, virtually it consisted only of almost single finely-dispersed carbon. Admixtures in the sample — gas constituents of carbonaceous matter (nitrogen, oxygen), the sulfur as well as very small amounts of earth silica and other silicates — constitute about 2.5 %.

The samples of massive and brecciated shungite rocks by their chemical composition were in content with design requirements. In the massive rock, there is — 31.4 % of carbonaceous matter, 65 % of earth silica, 2.6 %, of iron, in the brecciated rock, correspondingly, — 30.9, 64.5 и 2.4 %. % . As we see, these both samples are close by their chemical composition.

The third sample, of laminate shungite rocks, contained 85.3 % of carbonaceous matter and 5.4 % of earth silica. This sample in principle refers to the most rich shungite rocks.

On the example of rich laminate shungite rocks with large content of carbonaceous matter, it was supposed to define the presence of carbon allotropic configuration by way of fullerene, which crystal lattice structure has been already well-defined in Krasnoyarsk Research Center of High-technologies of Siberian State Aerospace Universrity named after academician Reshetnev M,F. and demonstrated in publication [Kashkin etc., 2000]. X-Ray picture of this structure is shown on Fig. 2.6

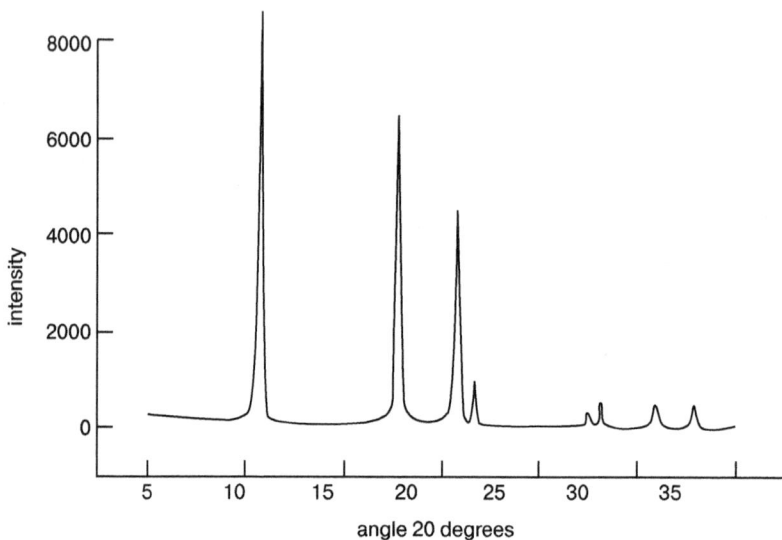

Fig. 2.6. X-Ray spectrum of fullerene-bearing soot — carbon globular structure with C_{60} formula

In the richest shungite rocks, an amorphous carbonaceous matter prevails. Judging by X-Ray, crystal structure in the laminate rock is available in small amount. These rocks can be widely used as a mineral fertilizer or as a coloring black pigment. Together, they may be added to more low-grade shungite rocks in the averaging process.

In the sample, which we investigated, of the rich laminate shungite matter, the presence of fullerene structural lines has not been noticed (Fig. 2.7a). Apparently, the fullerene matter amount is so little that used by us equipment DRON-3 does not allow to find it. Therefore, we think it is important to say that the existing opinion about positive fullerene role on the remedy influence on human organism has not been established.

Shown on Fig. 2.7a X-Ray picture of rich shungitous matter consists mainly of amorphous carbon with globular structure. The presence of small peaks on diagram points on the possibility in the rock of graphite, earth silica, aluminosilicate and other admixtures.

X-Rays of massive and brecciated rocks are approximately of one type (Fig. 2.7 b,c) but they significantly differ from rich laminate rocks. Besides the major amorphous matter, the silicates occur in these samples, Fig, 2.7 b,c. It has been stated by the way of obtained X-Ray spectra analysis (Table 2.3).

50

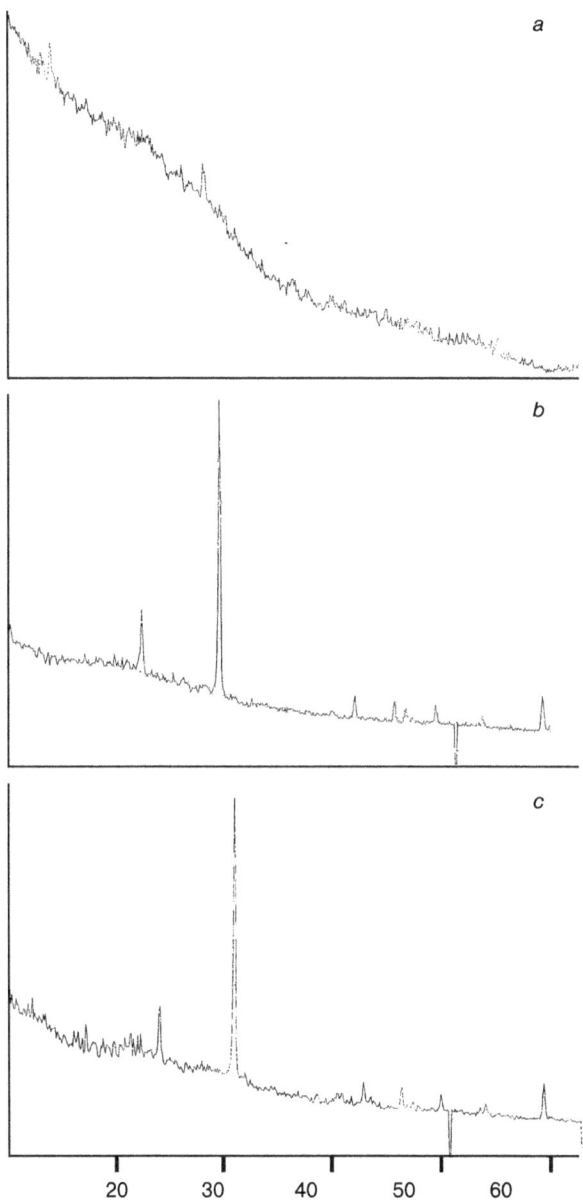

Fig. 2.7. X-Ray:
a — of rich shungite rock of laminate structure; *b* — of massive shungite rock;
c — of brecciated shungite rock

51

Table 2.3. Interpretation of X-Ray spectra

№	Sample 2				Sample 3				h	k	l
	Peak intensity	P	$10^4/d^2$ exp	$10^4/d^2$ calc	Peak intensity	d, Å	$10^4/d^2$ exp	$10^4/d^2$ calc			
1	midde	4.2625	550	551	middle	4.2549	552	552	1	0	2
2	strong	3.3490	892	893	strong.	3.3493	891	893	1	1	2
3	weak.	2.4574	1656	1653	weak.	2.4582	1655	1656	1	0	5
4	weak.	2.2825	1920	1918	weak.	2.2838	1917	1917	2	1	2
5	weak.	2.2412	1991	1994	–	–	–	–	1	1	5
6	weak.									0	4
7	weak.									1	4
8	weak.									1	5

As it is seen from Table 2.3, both samples have crystal substance in terms of silicate with tetragonal structure a = 5.411(6) Å, c = 13.81(2) Å.

In such a manner, investigated shungite rocks comprise mainly of amorphous carbonaceous matter with little admixture of earth silica and other over-crystallized mineral phases.

It's worth to outline that rich rocks in comparison with low-grade ones are characterized by more homogeneous material composition. In this connection, the necessity has appeared to classify all varieties of both rich and low-grade shungite rocks as far as the area of their supposed application, as practice has shown, is essentially different.

2.5. Explanation of Necessity of Complex Processing of All Shungite Rock Varieties

All shungite rock agglomerate across the deposit is rather equal, however, as it was mentioned above, on different deposit sections their chemical and mineral compositions as well structural features vary significantly. Hence, before to use shungite rocks in industrial conditions it is needed to classify them by the content of major components and physico-mechanical properties and on the basis of the classification to state the application sphere.

Such classification will much simplify the search of available customers for different products, including low-grade by carbonaceous matter ones, that will essentially expand the sphere of their possible usage,

It's worth to note that near major Zazhogino deposit, there are several, already investigated shungite rock sections with carbonaceous matter content from 2 to 15 %. In some samples, taken from the sections, richer shungite rocks present with carbonaceous matter content from 35 to 37 % that greatly complicates the areas of their application. Such drastic alternations of carbonaceous matter in shungite and shungitous rocks outline their big diversity that all the more so demands their classification. Mineral content and structure of shungite rocks in these sections are similar by their location. The uncovering level on exploited section "Mironovskiy-1" constitutes from 2 to 5 meters. The section is nearby major Zazhogino deposit. The predicted volume of shungite rock reserves on this section of 22 km^2 square is approximately 220−250 thousand tons. There is a license on industrial exploitation of section "Mironovskiy-1".

Before the beginning of this section exploitation, its detailed over-testing was pursued due to task LLC "MKK-Engennerring. The drill samples, taken from various drill depths, were analyzed on following chemical components: Al_2O_3, SiO_2, CaO, Fe_2O_3, C_{com}, S_{sulf}, SO_3, P_2O_5, loss on ignition. The total number of samples was 160. The results of samples chemical analysis, selected from different intervals from 3 drills, are shown in Table 2.4.

The chemical analysis confirmed that carbonaceous matter content in shungite rock is not large and alternates in average from 3 to 15 %. In some samples, it reached 30−35 %, however, there were only few samples, about 5 % of total number, with big content of carbonaceous matter. It is supposed that in the process of further research the number of rich rocks can increase in account of non-drilled cracks and single clusters.

Because first results of over-testing has showed that section "Mironovskiy-1" is presented mainly by low-grade by carbonaceous mater ettles, their application sphere has not been elucidated yet. With carbonaceous matter content in the rock of 3−15 % in average, the silica oxide alternates there from 40 to 70 %, the calcium oxide from 0.5 to 18.5 %.

The presence of calcium oxide big amount points on the carbonate basis of these rocks; the content of aluminum oxides in average is from 8 to 15 % that says about the availability of pelitic minerals in the rocks. In the process of additional rock enrichment by washing-off, one can partially wash out the pelitic minerals as well as other finely-dispersed mineral components. Such operation will allow to change physico-chemical properties of washed-off rocks and find the new sphere of their application.

The iron oxide content in a whole in all samples of section "Mironovskiy-1" changes from 1.97 to 10.1 % that points on big amount of pyrite, magnetite, hematite and other ferrous minerals in comparison with main deposit rocks.

Table 2.4. **Chemical analysis of samples taken from section "Mironovskiy-1"**
(according to VIMS data)

№ sample	Content, % mass.									
	Al_2O_3	SiO_2	P_2O_5	CaO	Fe_2O_{3com}	C_{com}	ignit loss	CO_2	S_{com}	SO_3
C-370–1	10.4	55.1	0.07	0.58	5.66	14.0	20.1	<0.10	2.41	0.14
C-370–2	10.4	55.9	0.09	1.63	10.1	6.00	9.42	0.76	0.88	0.09
C-370–3	12.2	57.1	0.09	1.83	8.38	4.24	8.96	0.54	1.56	0.11
C-370–4	6.25	39.5	0.12	18.5	8.79	4.29	16.1	13.6	1.56	0.18
C-370–5	10.8	64.3	0.09	1.92	5.71	1.50	4.91	0.30	0.79	0.14
C-370–6	15.1	59.1	0.11	1.08	4.09	4.62	7.59	0.14	0.66	0.08
C-370–7	11.3	63.8	0.11	1.76	4.70	4.54	8.23	0.35	0.89	0.10
C-370–8	9.07	57.4	0.10	3.24	7.85	8.27	12.4	1.18	0.80	0.22
C-370–9	8.20	38.9	0.14	14.0	9.39	11.0	19.3	8.42	1.67	0.20
C-370–10	11.1	51.7	0.16	5.50	5.43	6.07	11.7	2.50	0.76	0.14
C-371–1	5.56	60.7	0.05	0.21	5.08	14.7	20.6	<0.10	3.60	0.20
C-371–2	7.32	70.1	0.07	<0.10	4.51	8.37	12.3	<0.10	3.18	0.28
C-371–3	10.4	65.5	0.10	0.12	5.65	4.95	11.4	0.23	4.01	0.14
C-371–4	9.09	68.4	0.07	0.15	3.87	4.33	13.0	<0.10	3.20	0.17
C-371–5	8.84	67.4	0.08	1.79	4.44	5.62	9.97	1.07	3.09	0.11
C-371–6	9.70	68.7	0.08	0.15	4.90	5.83	9.71	<0.10	3.63	0.19
C-376–3	4.18	48.3	0.05	0.15	1.97	37.5	42.6	<0.10	0.92	0.14
C-376–4	4.05	47.9	0.04	0.17	2.38	37.6	42.9	<0.10	1.94	0.13
C-376–5	4.11	48.6	0.05	0.16	1.97	37.1	42.0	<0.10	2.12	0.15
C-376–6	4.04	50.1	0.05	0.13	2.57	34.8	38.7	<0.10	1.12	0.04
C-376–7	9.47	52.0	0.07	3.50	5.53	8.10	20.8	16.3	0.95	0.27
C-376–8	13.7	58.0	0.11	2.07	5.36	3.15	7.52	1.11	1.91	0.15
C-376–9	13.1	60.7	0.11	0.99	4.86	4.09	7.92	0.30	3.12	0.09
C-376–10	13.7	60.9	0.16	0.90	4.79	4.61	8.51	<0.10	2.15	0.12
C-376–11	10.8	67.0	0.10	0.53	3.94	4.86	7.96	<0.10	1.95	0.08
C-376–12	9.70	63.8	0.09	3.63	5.16	2.11	6.11	2.62	2.11	0.22

During the analysis, it has been figured out that in all studied samples in shungite rock matrix, the sulfurous compounds exist, apparently, mainly sulfides. The content of common sulfur in the samples alternates from 0.66 to 4.01 %, and the ratio of sulfurous compounds is from 0.5 to 1 %. However, in some shungite ettle samples, the ratio of sulfurous compounds reaches 3 %.

The low-grade shungite rocks with sulfurous compound content below 1 % have pH of about 6, and while the grow in their ratio till 3 %, the pH does not lower below 5.5.

The pursued chemical analyses of shungite rocks of section "Mironovskiy-1" demonstrate that they are much more deficient in carbonaceous matter that the rocks of main Zazhogino deposit. The application area of the ettles from section "Mironovskiy-1" has not been sufficiently substantiated yet. Therefore, additional conductance of industrial tests with the purpose to establish the technological parameters and the areas of supposed application of deficient in carbonaceous matter rocks will be needed. This will allow to state and find practical sphere of ettles application which chemical composition drastically differ from ettles of major deposit. Such ettles need additional enrichment before their usage.

Starting from here, one should study all complex of their technological properties with the account of their mineral, chemical compositions and structural peculiarities that will allow to find for these rocks the new practical areas of application in different production spheres with usage of concentration (enrichment) methods or without them.

To substantiate in more detailed manner the new areas of application of low- and high-grade shungite rocks it is necessary to classify them by chemical and physical properties and, on that classification basis, to shed a light on novel predicted spheres of their application, confirming them experimentally.

2.6. Shungite rock classification by carbonaceous matter and mineral admixture chemical content

It is stated that carbonaceous matter content in shungite rocks alternates from 0.5 to 98 % [Borisov, 1956]. It was based on the chemical analysis of large number of samples, selected from different Zazhogino deposit sections.

Structural physico-chemical features and chemical content of shungite ettles also drastically varies that is, surely, changes the sphere of their application.

Therefore, the classification of all shungite rock varieties into several groups, their division, first of all, by carbonaceous matter content is needed on the investigation primary stage. For the first time such classification was worked out by Borisov P.A. yet in 1932 (in Table 1.1).

According to this classification, the shungite rocks were divided only by carbonaceous matter and earth silica content, while the other shungite rock constituents were not taken into account, and their application spheres were not specified.

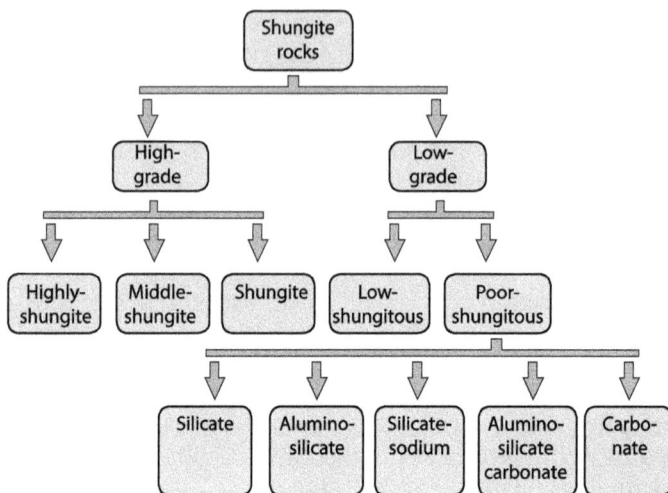

Fig. 2.8. Scheme of technological classification of shungite rocks by carbonaceous matter content and admixture chemical composition

To characterize already known and perspective varieties of shungite rocks and specify their probable application sphere, their new classification is proposed with the emphasis not only on the carbonaceous matter content percentage but also on the silica oxide, calcium, carbonates, aluminosilicates and other silica-bearing compounds. Each type of these ettles has their sphere of application in different industry branches.

Carried out technological classification of major groups of shungite and shungitous rock groups by the carbonaceous matter content and the chemical composition of admixtures is presented on Table 2.2. The classification schematic generalized picture is shown on Fig. 2.8.

Following the new classification, we want to give the technological evaluation of all five shungite rock groups separately, outlining herewith the possible perspective sphere of their application with account of chemical content and technology properties.

The richest by carbonaceous matter content (65–98 %) are highly-shungite rocks. Their volume in a whole along Zazhogino deposit is not large and they are situated in the deposit major depth in a kind of separate zones.

Such rocks can be used as technological fuel, mineral fertilizer and pigment for paint production as well as in industry where electricity-conductive materials are in need. Industrial trials on the usage of such high-carbonaceous rocks were being pursued multiple times, however, their systemization by physico-chemical properties has not been made.

Table 2.5. Technological classification of shungite rock major groups by carbonaceous matter and admixture chemical composition

Ettle types	Group name	Chemical composition of major components, %					Application sphere-production
		C	SiO_2	Al_2O_3	Na_2O	CaO	
Rich by carbon							
I	Highly-shungite	65–98	10–20	0.1–0.5	0.1–0.5	0.5–1.0	Mineral paint, fertilizer, activated shungite product
II	Middle-shungite	45–65	15–40	0.5–1.0	0.1–1.0	0.5–1.0	Calcium carbide, fertilizer
III	Shungite	25–45	35–65	1.0–3.0	0.1–1.0	0.5–1.0	Rubber technica, pyrometallurgy construction industry, chemical industry
Low-grade by carbon							
IV	Low-shungitous	15–25	40–75	7.0–9.0	0.1–1.0	0.1–1.0	Construction materials, warmer, brick, plaster,chemical industry
V	Poor-shungitous: silicate	0.5–15	70–95	1.0–5.0	0.1–0.5	0.1–1.0	Iron and non-iron meatllurgy, construction industry
	Aluminos-ilicate	0.5–15	30–45	10–20	0.5–3.0	1.0–3.0	Construction industry: shungisite, road metal, filler for light concrete
	Silicate-sodium	0.5–15	30–60	10–15	5.0–30	0.1–0.5	Chemical industry: sodium silicate, calcium carbide, calcium chloratum
	Aluminos-ilicate-car-bonate	0.5–15	30–60	10–20	1.0–5.0	15–25	Construction industry: cement, road metal
	Carbonate	0.5–15	20–40	5–10	0.5–1.0	20–60	Chemical industry: calcium chloratum and metal caclium, lime

The positive results were obtained while the usage of these rocks as a technological fuel. The ash output while the burning of ettles with high carbonaceous matter ratio is small, but energy value is rather high. Such shungite rock type reminds regular antracite. However, their industrial introduction has not been accomplished because these rocks location across the deposit has not been defined, and it has been impossible to exploit the deposit by the way of spot getters of rich sections.

The special application area can be for shungite rocks of the second type with carbonaceous matter content from 45 to 65 %. These groups shungite rocks can be also used according to their technological properties like the rocks of the first type. Although the additional investigations to figure out their advantages and drawbacks in each particular case should be conducted. These works should be pursued after the shut of the being constructed industrial facility. Statistically substantiated data for this rock group may be obtained only after long interaction between supplier and customer but this is a matter of time.

The largest interest across the deposit represents shungite rocks of the third type, wherein the carbonaceous matter content is 25–45 %, the earth silica — 35–65 %, the silicate minerals — 5–10 %. The total ratio of such rocks across the deposit is over 45 % [Borisov, 1956]. This shungite rock type is accepted as the most optimal for practical use, and constructed enterprise is oriented towards their chemical composition.

This shungite rock variety was used by VIMS for basic industrial introduction, herewith physico-chemical properties of all products had been investigated, as well as the spheres of application in different industrial branches (construction, metallurgy. chemical, industrial-rubber, paints and lacquers) had been established. Mineral, chemical compositions and properties of such shungite rocks are considered in details in Chapter 1.

Low-shungitous rocks with carbonaceous matter ratio of 15–25 % have separate sphere of application. Such rocks application sphere has not been enough stated yet, for this, special investigations are at demand. These rocks on the first stage of industrial introduction might be used by the way of averaging with richer middle-shungitous rocks. As a result of such mixing one can easily get the shungite rock with the Project content of carbonaceous matter 25–45 %. Thereat the total volume of rocks with the Project content of carbonaceous matter grows up more than 2 times.

As it has been already mentioned, poor by carbonaceous matter rocks with carbonaceous mater content from 0.5 to 15 % are also developed in the deposit. The total volume of such rocks across the deposit is rather high, and there are many of their varieties: silicate, aluminosilicate,

silicate-sodium, aluminosilicate-carbonate, carbonate. They are widely spread in the frames of major deposit and especially on its adjoined sections. Where all aforesaid varieties of poor shungite rocks present. Their technological properties have been studied poorly, and the application sphere has not been enough substantiated yet. Such rocks trials should be expanded. Shungitous rocks have big number of additional mineral substituents by way of tuff particles, carbonates, sulfides, silicates, aluminosilicates. This rock group, which is poor by the carbonaceous matter content, has been badly studied in technological terms, its application area has not been fully determined, their physico-chemical features demand additional study with account of bedding, location and possible getter conditions.

Several variants are supposed for the usage of these rocks as a raw for cement, keramzit, liquid glass and other constructing materials production. Poor-shungitous rock mineral composition and chemical properties allow to use them as a raw for darkening glass and sodium silicate production. These rocks must be exposed to additional enrichment before their usage by either gravitational, pyrometallurgical, or hydrometallurgical methods.

The choice of ways and methods of the poor shungitous rock concentration depends on their chemical composition as well as needed quality of the concentrates. The promising method for the enrichment is thermal, which allows to transfer, by the way of thermal processing, the rock particular minerals from one state to another by burn. As well as the chemical solvation of particular chemical components, for example, of sulfides at definite pH values, is also promising.

Poor shungitous rocks, bearing sodium and potassium silicates, can be also used as a raw while the production of liquid glass, rather scarce chemical material, which is widely used during the flotation of rare and non-ferrous metal ores of various category as well as in the mining industry.

Poor shungitous rocks find their application in construction industry, for example, while the making of cement solution, shungite concrete, shungizite, special decorating shungite solutions, creating the screening surfaces in the special purposes buildings. They can be used as a substitute for road metal in the shungite concrete production. Such concrete will have low relative weight and possess low thermal conductivity that is rather vital in residential building construction in North areas. Poor shungitous rocks can be used also while the production of thermal-protecting floors in industrial and residential buildings.

The application sphere of poor shungitous rocks with large content of pelitic minerals, field spars, other aluminosilicates and carbonates has not been sufficiently substantiated yet. The role of these rocks in the industrial introduction process has not been figured out yet, and technological evaluation has not been done. The silica oxide ratio in such rocks reaches 40—50 % and more. As we suppose, they can be used in pyrometallurgical industry while the production of raw iron and ferro-alloys as well as in non-ferrous metallurgy. However, the industrial tests with such shungite rocks type has not been pursued yet. After the industrial enterprise shut, such tests should be embodied in the series of big plants of ferrous and non-ferrous metallurgy with the purpose to establish the rock application sphere and their common technological evaluation.

The application spheres of poor-shungitous rocks, enriched with carbonates and aluminosilicates, and the rocks, enriched with silicates, differ much.

In all the varieties of poor-shungitous and low-shungitous rocks, there is a big amount of molybdenum, strontium, vanadium, selenium, nicol which can cause significant impact on physico-chemical processes of fusion, intensifying the redox effects of pyrometallurgical processes. So, the listed elements play catalyzing role in these reactions. Such minerals, which contain iron, coper, nicol and others, can also play catalyzing role. The action mechanism of such catalysators on the redox processes has not been learned yet; just several suppositions exist substantiating their effectiveness. The problem of fore-enrichment of shungitous rocks has not been elucidated yet for the reason of customers lack despite all attempts of some institutes. Therefore such questions settlement is possible only after the finish of the construction of the plant on shungite rock production and the availability of supposed customers.

Taking into account the big variety of shungite and shungitous rocks, it is necessary, as far as their getter works and processing proceeds, to begin the research on the opening of some new possible additional technological applications.

Thus, following this classification, all shungite, low-shungitous and poor-shungitous rocks can be used in various industrial branches. For this, it is necessary during the construction and introduction to new projected production, primarily to work out there the technology of possible application of different shungite rock varieties, first of all, on the factually optimal shungite rock, that is of first sort, and then, after the understanding of customable properties, to begin works to shed a light on new areas of application of other varieties. This is, firstly, is very vital for high-shungite, middle-shungite and shungite rocks and then for low-shungitous and poor-shungitous ones. All the works in this direction has just begun.

2.7. Theoretical Statement of Shungite Rock Technological Classification

It is clear, because of the big variety of afore-described shungite and shungitous rocks, every kind has differing, rather specific properties and, surely, will have its own application sphere. Hereat they must be necessarily divided into several groups. Then every group should be theoretically stated, outlining herewith the properties and possibilities to use.

If to consider shungite rocks of Zazhogino deposit following the chemical-genetical classification of Galdobina L.P. (in Fig. 1.3), presented by way of triangle, the optimal calculous chemical content of shungite rocks obtained in the project with carbonaceous matter content of 25–45 % will be in the center of this equilateral triangle. Sizing the circle to fit the triangle, we will get inner contour which contains shungite rocks of II and III groups. The volume of this square, apparently, will be 75–80 % of total Zazhogino deposit volume (Fig. 2.9).

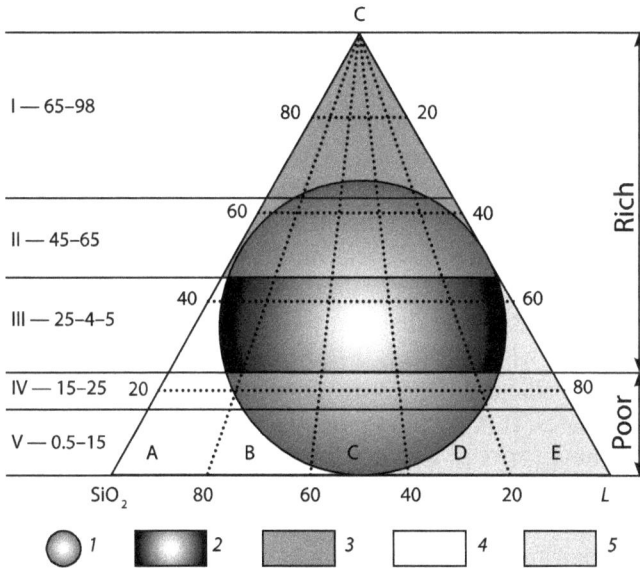

Fig. 2.9. Diagram of technological classification of rich and poor shungite rocks with account of their chemical content: *1* — major volume of shungite rocks; *2* — projected rocks; *3* — shungite rocks with large content of carbonaceous matter (groups I–II); *4* — shungite rocks with low content of carbonaceous matter and high earth silica content; *5* — shungite rocks (group IV–V) with low carbonaceous matter content and high carbonates content

The square of fitted circle in the considered triangle represents the average chemical composition of shungite rocks across the deposit. The industrial production has been projected for these rocks and all necessary industrial trials have been conducted. Namely, they will be used in different industrial branches along the set tasks of the plant.

The rocks, not got into the main group on diagram, are presented by three sections of the triangle, upper section reflects the rich shungite rocks, located of the circle outer space, with carbonaceous matter content from 65 to 98 %, and left and right sections — are poor shungite rocks with thecarbonaceous matter content from от 0.5 до 15 %; shungite rocks application areas, situated at different parts of diagram, differ essentially.

To increase the common volume of processed rocks with optimal content of carbonaceous matter of 25–45 % it is necessary to conduct their additional averaging that is easily to make using the triangle diagram.

The shungite rocks with high carbonaceous matter content must be mixed with shungitous ones with carbonaceous matter low content before their processing. The composition of the mixture of these two ettles will be equal to average chemical content of major deposit which composition is supposed in technological regulations. Such mixture will be inside the circle fitted into the triangle of technological classification. The offered averaging technology will allow to expand the volume of shungite rock processing significantly.

To divide the shungitous rocks on the triangle diagram by their mineral varieties (silicate, aluminosilicate and others) let draw the lines from top point of the triangle diagram to its bottom: A — the area of poor rocks enriched with silicates, B — with aluminosilicates, C — with sodium aluminosilicates. D — with aluminosilicates and carbonates, E — with carbonates. It is worth to note that major volume of shungite rocks contains these varieties also but their quantity is mutually decreases as far as carbonaceous matter content increases.

Thus, the considered triangle diagram has allowed to divide all the varities of rich and poor shungite rocks into 5 additional varieties. As it is seen from the diagram, on the left, there are silicate rocks with carbonaceous matter low content, on the right, carbonate rocks with very little admixture of earth silica.

The special status is for shungitous rocks bearing small amount of carbonaceous matter and silica oxide. On the considered diagram, this is the right side of the triangle. Low-shungitous and poor-shungitous rocks (groups IV and V) are characterized by high content of carbonates, aluminosilicates, field spars. Tuffs and tuff siltstones might be referred to them. Their application sphere and major features are rather specific and scarcely studied. Additional expenditures to conduct such investigations are needed. These rock volume across the deposit is small. However, there is rather much of similar rocks if to take them altogether with row of little deposits around major Zazhogino deposit.

As we see according to this classification the total volume of all Zazhogino deposit rocks which might be used in industry, is rather high — over 90 %. For full usage of low-shungitous and poor-shungitous rocks, it is needed to expand the experimental jobs to establish their application spheres. Evidently, they must be exposed to additional concentration before the usage. The enrichment degree and the quality of thereat obtained products will depend on technical conditions (TC) of new customers.

2.8. Development of Methods of Technological Evaluation of Shungite Rocks

It is shown above that shungite rock chemical content, especially that of poor by carbonaceous matter, is rather varying. Each variety has its own, rather specific application area. As it was already mentioned, the poor shungite ettles can be divided into several types: silicate, aluminosilicate, silicate-sodium, aluminosilicate-carbonate carbonate and also mixed.

All these varieties have their own rather specific properties, and, undoubtedly, their application sphere will significantly differ from application sphere of standard rich shungite rocks. Although it has not been stated yet, it is worth to note that shungite rocks, depending on their characteristics, in future, must be exposed to additional concentration with account of supposed clients.

The technology of shungite rock enrichment for each kind should be additionally carried out with account of possible customers. Therefore, to assess the specific properties of these ettles the special methodology should be created with the purpose to learn the supposed application spheres of these kinds.

In present, while the construction of plants on shungite rock processing such methodology is extremely vital. The classification of poor shungite rocks by chemical content and technological properties must be held by principle of minimal self-cost and minimal level of sorting.

The division process can be conducted either at the bedding place or industrial facility territory. In the evaluation of poorer ettles it is necessary to take into account their volumes, possible reserves of their mixing with other kinds and planned perspectives for their further application.

According to the worked-out technological classification, all shungite rocks are divided into two major kinds: rich and poor. The main evaluation criteria for rich ones is the presence of the carbonaceous matter, which content according to, one can immediately define which shungite rock kind the considered enterprise delivery for processing might be referred to. For example, it is stated due to the chemical analysis that carbonaceous matter ratio in the rock is 75 %. According to the chemical technological classification and presented triangle diagram the rock types and sphere of their application can be determined.

The methodology of technological evaluation has been carried out for each kind of shungite and shungitous rocks. The essence of the methodology is in the following. As it has been noted, all shungite rocks are divided into rich and poor in accordance with developed technological classification. For rich ones, the major evaluation criterium is the presence of carbonaceous matter, which content according to, one can headily define the considered delivery — which shungite rocks kind the delivery for the processing can be referred to.

Thus, the shungite rocks delivered to the plant before the processing should be analyzed on carbonaceous matter content which content percentage the rock type is defined on: rich or poor.

If the carbonaceous matter content is higher than 45−50 % the application sphere of such rich shungite rocks is being immediately disclosed.

In the Table 2.6, the main parameters for all shungite rocks kinds, which are necessary to be determined with the help of chemical analyses, are presented.

The application sphere of major shungite rocks with carbonaceous matter content of 25−45 % is established similarly; if carbonaceous matter content is less than 25 % the rock evaluation must be held by other methodology, herewith, SiO_2, Al_2O_3, CaO, Na_2O, K_2O are determined additionally. According to the ratio of the latter compounds and technological diagram C-Si-L, the shungite rock kinds are outlined, their possible application sphere is determined.

Obtaining the analysis results the specialists should separate these rocks from major mass. Such rock sorting would significantly simplify their assessment and would give the possibility to systemize the processing regime for all rock varieties.

Table 2.6. **Shungite rock types assessment according to carbon and chemical composition**

Rock type		Carbon content, %	\multicolumn{2}{c}{Analyzed chemical composition}
Class	Subclass		C	SiO_2, CaO, Al_2O_3, Na_2O, K_2O
Rich	High shungite	65–98	+	–
	Middle-shungute	45–65	+	–
	Shungite	25–45	+	–
Poor	Small shungitous	15–25	+	+
	Poor shungitous			
	silicate	0.5–15	+	+
	aluminosilicates	0.5–15	+	+
	Silicate-sodium	0.5–15	+	+
	Alumino-silicate-carbonate	0.5–15	+	+
	Carbonate	0.5–15	+	+

Note. + — analysis is pursued;
– — analysis is not pursued.

Conclusions

1. As a result of structural, physico-chemical and mineral characteristics of all shungite rock kinds it has been stated that they mainly consist of organic non-crystal carbonaceous matter which have hollow globular skeleton consisting of 60 carbon atoms; earth silica in finely dispersed state is evenly spread inside the structure.

2. It is supposed that the major component — the shungite rock carbonaceous matter — has been formed from organic sediments of plant origin of sapropel type, which were affected by metamorphization with the formation of big number of micro- and macro- grikes, alvelouses and inner hollows. These grikes were being filled with finely-dispersed earth silica of different structural modifications, iron-containing minerals in the way of pyrite, hematite, magnetite, silicates as well as carbonates, aluminosilicates, sulfides. Shungite rocks contain wide complex of micro elements. The technical specificities of shungite rocks are defined by their complex mineral and chemical composition.

3. Specially pursued detailed analysis of pure carbonaceous matter from different Zazhogino deposit sites has shown that in all sections of the deposit, the carbonaceous matter contains volatile gas components — oxygen, nitrogen, hydrogen, sulfur gases. Sulfur gases

percentage is small and, in average, is 1.5–2.5 %. While the heating the chemical elements quickly fly and, at 100–150 C temperature, completely escape. Sulfur and oxygen play special role in the chemical interaction processes and redox processes. It is impossible to get chemically pure carbon because of the presence of gas element admixtures inside carbonaceous matter. Usually, the maximal ratio of carbonaceous matter which can be extracted in concentrates, doesn't exceed 98 %.

4. The light has been shed on that sulfide veins and concretions, some varieties of finely-dispersed earth silica of different polymorphic modifications are available in shungite rocks besides major shungite matter matrix consisting of carbon and earth silica. The form & structure of finely dispersed earth silica may be different, the particle coarseness — is from 1 to 100 microns.

5. The pursued DTA analysis has shown that all shungite rock varieties much differ between each other despite their close properties. Therefore, it is necessary to pursue the classification of obtained products with account of customer needs in the process of their industrial usage.

6. X-Ray analysis demonstrates that finely dispersed silica and other mineral admixtures present inside amorphous carbonaceous matter.

7. The wide spectrum of shungite and shungitous rock varieties has been determined by the carbonaceous matter, earth silica and other components content in there. To specify such rock sphere of application, their principally new classification has been carried out. It has been demonstrated that all rocks due to carbonaceous matter content might be divided into rich and poor.

8. The rich by carbonaceous matter (over 25 %) shungite rocks can be divided into three kinds with account of the sphere of their application by new classification: high-shungite, middle-shungite and shungite. High-shungite rocks with carbonaceous matter content over 65 % may be used as power fuel and mineral fertilizer, middle-shungite rocks with carbonaceous matter content from 45 to 65 % should be used as a raw for fillers for paint production as thermal insulation while building of the walls which possess the screening features in special institutions. These ettles are recommended to be used as an additive to small-shungitous rocks while their mixing with the purpose of the averaging.

9. And, finally, major and more promising consumption sphere of shungite rocks with carbonaceous matter content of 25–45 % — is industrial rubber industry wherein technical carbon (carbon coal)

and synthetic earth silica (white coal) are substituted. These products are widely used in tyre and elastomer production. Besides, these rock variety can be also widely used in metallurgical production and in the production of construction materials as it has been proved by industrial practice.

10. As a result of fulfilled research, it has been established that poor by carbonaceous matter shungite rocks (0.5–15 %) can be also widely used in different industry spheres, including construction, mining-enriching, chemical, metallurgical. It has been stated that disclosed varieties of poor-shungitous ettles: silicate, aluminosilicate, silicate-sodium, aluminosilicate-carbonate, carbonate and others can be successfully applied as original raw in different production branches.

11. Triangle diagram has been carried out and built on the basis of physico-chemical peculiarities of Zazhogino deposit rocks. This diagram allows to distinguish the different varieties of shungite rocks with account of carbonaceous matter, earth silica and other component — as aluminosilicates, carbonates etc. — content.

12. Special methodology on the evaluation all shungite rock varieties has been worked out in accordance with proposed diagram. Thus, for rich rocks, one should define just the carbonaceous matter content and to establish on its presence basis the area of their possible usage. For the rocks, poor by carbonaceous matter content, it is needed to pursue additionally the analysis on silica oxide, aluminum. calcium, sodium, potassium content besides the carbon definition. The variety of the rocks and sphere of their possible usage is determined by these oxides quantity basing on the proposed diagram. The technology of distinguishing and classification for each variety of poor and rich rocks is identical, however, for poor rocks with variable chemical composition, additional special research is needed to specify novel promising spheres of their application.

Chapter 3
TRANSFORMATIONAL TECHNOLOGIES
OF SHUNGITE ROCK PROCESSING

3.1. Substantiation of the Need to Produce Shungite Products with Pre-Determined Characteristics

As it is well known, the shungite rock processing technology is very simple; it includes two-stage grinding, size specification of the fragmented product and drying. Also, the part of the fragmented product is a subject to all-sliming. Due to its simplicity such technology of shungite rock processing does not always comply with the specifications of consumers, in particular, speaking about quality which is usually assessed by the content of the carbonaceous matter and silica in the original rock.

Besides, the original rock has many other components (e.g., iron or copper sulphides etc.) which form the acidic medium. It has negative impact on further application when manufacturing of new technology products.

Acidic shungite products in industrial-rubber goods increase the vulcanization period. Thus, the vulcanization period increased approximately by 6 % when white carbon had been replaced with a shungite product. For decreasing the acidic features of the applied shungite fillers we have tested alkaline fillers — soda, calcium carbonate and talcum powder. But the additional fillers-modifiers can also influence the quality of protector rubber.

The acid medium is formed during grinding, size specification and all-sliming. Specifically, pH of the aqueous recover of shungite products is 3–6. The acidic condition increases due to the content of elemental sulphur and sulphides in the rock. During the processing these components oxidize quickly as affected by the ambient air.

The acid medium which was formed during the processing negatively influences the quality of the obtained product used as a filler for industrial-rubber goods. The hydrogoethite, magnetic iron oxide and other ferrous minerals in the rock also can influence the quality of the shungite product. Without doubt, carbonate, silicious and aluminosilicate components, presented in the bulk mass of shungitous rock, also have a negative impact, but their role is underexplored.

So according to the classification developed for complex processing of shungite rock, each group of rocks shall find its consumers. Thus, the majority of shungite rock types can be applied for producing of industrial-rubber goods, while other low-grade types — can't, due to the high content of admixtures. In the case of using them in industrial-rubber goods the rubber quality is reduced significantly.

These types of rocks should be used in other spheres where silicates, carbonates, aluminosilicates etc. are allowed by technical conditions. The problem is completely unexplored in the practice of using shungite rock both in the technological and physiochemical senses. Meanwhile the general utility of all low-grade types of shungite products will depend on and will be determined on the basis of the knowledge about them.

Without doubt, carbonates such as dolomite, tiff, feldspars and other components of shungite rocks can also cause negative impact on the using of low-grade shungite rock. It is not recommended to use shungite rocks with carbonate content which is higher than 2.5 % in industrial-rubber goods.

Carbonate inclusions in shungite rocks may lead to reduction of rubber performance. So, low-grade shungitous carbonate rocks should be revealed during technological processing and if possible separated tfrom the main bulk of processed shungite rocks,. During the processing, low-grade shungitous carbonate rocks should be neutralized or transformed into other forms, thus, developing the technology of their possible removal or dissolution.

During this process for neutralizing the acidic medium, the carbonate rocks can be added to high-grade rock in small quantities. However, these problems require a thorough analysis after industrial testing of rocks and finding the percentage of carbonate additive. Such additive volume shall not exceed 2.5 %.

Silicate minerals also cause a negative impact on processing. Almost all types of shungite rocks contain them, especially low-grade ones.

The considered criteria should be strictly controlled during the usage of shungite rock for full exclusion of negative impact on rubber processing technology. Due to potential negative impact of some components of shungite rocks, the processing flow chart shall be corrected additionally to exclude the potential variants of their impact. Blending (averaging) can be used for such correction, i.e. addition of low-grade rock to high-grade one for changing the quality of the original due to mixing rocks of different chemical composition.

Mechanical and chemical leaching is of big interest for changing of the qualitative composition of the original shungite products, so same as the transformation of minerals from one form to another by chemical dissolution or mechano-chemical washing out of some components. This allows improving the features of the obtained products.

Thus, most sulphides and organic sulphur can be removed by simple mechanical leaching without the addition of chemical reagents. The process of sulphide leaching can be catalyzed by adding a small quantity of ferrous sulphate decreasing pH to the optimal value (3–3.8). The consumption of the ferrous sulphate in average shall not exceed 0.5 kg per ton of the processed product. This figure is based on the laboratory experiments.

Carbonate minerals in low-grade shungite products can be dissolved by treating them with chlorohydric acid solution (concentration — 10–15 %). The reaction of carbonate leaching with chlorohydric acid is rather simple:

$$CaCO_3 + 2HCl = CaCl_2 + H_2O + CO_2\uparrow;$$
$$MgCO_3 + 2HCl = MgCl_2 + H_2O + CO_2\uparrow.$$

The original liquid chlorides can be evaporated from the water suspension and carbon dioxide is removed through the ambient. This process allows removing almost of all carbonates from the shungite rock.

The calcium chloride solution formed by carbonates dissolving can be boiled out in a flash drier for getting solid calcium chloride dihydrate ($CaCl_2 \cdot 2H_2O$) which can be filtered and used for the industrial needs.

The filtered residue of shungite rocks without carbonates will have completely different technological properties if compared with the original shungite rock.

It is necessary to leach carbonate compounds from shungite rocks due to rather thin interpenetration of the former in the carbonaceous matrix. This makes it impossible to separate them using gravitational and flotation methods. Carbonate compounds shall be leached in a way to manufacture then new chemical compounds during chemical treatment and use them as efficient substances for industrial needs after dissolving and separation from the main bulk of rocks. For this purpose, the carbonates were dissolved not only in chlorohydric acid, but also in salmiac.

As a result, we obtained calcium chloride and gases such as NH_3 and CO_2 according to the following reaction:

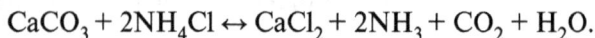

$$CaCO_3 + 2NH_4Cl \leftrightarrow CaCl_2 + 2NH_3 + CO_2 + H_2O.$$

After evaporating in the flash drier we manufacture $CaCl_2 \cdot 2H_2O$ from calcium chloride which can be used itself for its intended purposes. According to the laboratory experiments, reagent consumption is within the stoichiometric standard. Calcium carbonate dissolves quickly providing that the particle coarseness is less than 1 mm. If the temperature rises, the dissolution is catalyzed even in case of small concentration of solvents (5−8 %), Fig. 3.1, 3.2.

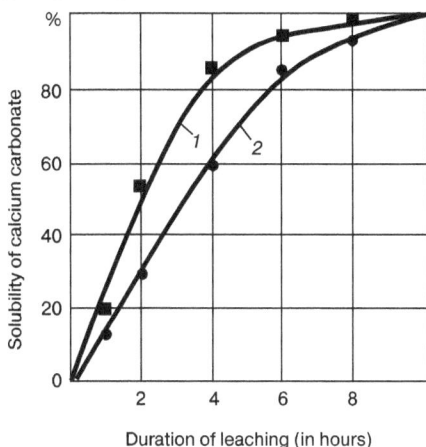

Fig. 3.1. The kinetics of calcium carbonate dissolving in chlorohydric acid and salmiac:
1 — 10 % solution of HCl; *2* — 10 % solution of NH_4Cl

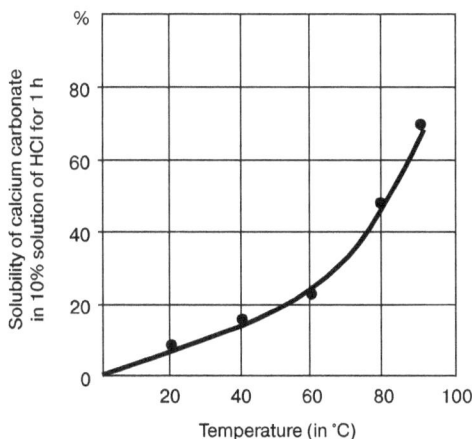

Fig. 3.2. The speed of calcium carbonate dissolution in chlorohydric acid depending on the temperature

71

Table. 3.1. **Types of mineral components and their possible solvents**

Components	Solvents
Carbonaceous	Organic solvents under pressure
Silicious	Fluohydric acid, alkali aqueous solutions
Ferriferous	Aqueous solutions of inorganic acids
Sulphur and sulphides	Carbon sulfide, acids, salts of inorganic acids
Carbonates (limestone, dolomite rock, chalybite)	Chlorohydric acid solutions, acid sulphate salts, sodium sulphide
Aluminic	Aqueous solutions of acids and alkali, salts of inorganic acids and bases
Natural salts	Water

It's shown that it's possible to remove carbonates from shungite rocks by burning under temperature of 800–900 °C. A burned carbonate shungite rock is a good raw material for lime.

In fact, all mineral components of shungite rocks can be easily dissolved. This changes their chemical composition and quality of the obtained shungite products depending on dissolution of the certain chemical components of the main matrix of the carbonaceous matter. For proving the above-given statement let us consider the solubility of chemical components of shungite rock. Main ways of dissolving harmful components in shungite rocks are given in Table 3.1.

As seen from the table, all mineral components of the shungite rock matrix can be dissolved and recovered from the matrix of the processed rock. The conditions of dissolution depend on the solvent composition, concentration, pressure, temperature and the oxygen amount in the dissolved medium.

As for the chemical composition, sulphur and sulphides cause the most technological harm for the shungite rock matrix. During processing they oxidize under the ambient air and increase the acidity of the obtained products. Specifically, pH is reduced to 3.5–4 depending on the content of these products. Transformation of certain mineral phases leads to changing the matrix structure. Obviously, this changes the physical-chemical characteristics of the original products. The intensity of mineral phase transforming depends on the time, PT-conditions, solvent concentration, coarseness of dissolved products and the amount of the air supplied for oxygenation of the suspension.

In some cases of shungite rock application, even silica and silicate minerals may negatively influence technological properties of the products. The process of controlling of their coarseness shall be strictly regulated and improved during the industrial manufacturing of shungite products including from shungite rocks with pre-defined negative characteristics.

All the above-given should be thoroughly regulated when manufacturing shungite products. Statistical data should be accumulated and the technological process shall be controlled, especially the fineness of grind. The allowable content of large- and small-sized particles in the product, obtained for application, should be specified in advance. It's very important for supplying the products to industrial-rubber goods plants. So for manufacturing of shungite products it is important to know the best composition of the original raw materials not only by ferriferous minerals, but also by presence of silica and other silicates, especially considering their coarseness and structural properties; larger silicate minerals with other structures can cause a significant negative impact within using shungite products.

When supplying the shungite products to the processing plant, one should deal very cautiously with the technological aspects as just slight exceeding of the technological rules not only in the chemical composition, but also in particle coarseness (especially that of silicate ones) may lead to negative results when using them.

To avoid such faults when applying shungite products, especially in industrial-rubber goods, once should consider the ratio of using shungite fillers (SF) instead of technical carbon (Π-900). This will allow no harming of the vulcanization process on the initial stage of the adopting of shungite substitutes. Firstly, only a part of technical carbon should be replaced by shungite products. Thus, at the beginning of the industrial implementation only 20 % of technical carbon can be replaced by shungite products. This should be controlled strictly.

The produced tires with shungite filler shall be tested under different road conditions both in summer and in winter.

Further increase of the percentage of shungite products in industrial-rubber products may be done only after positive results of the previous industrial testing. After the testing, the percentage of the shungite filler increase in the industrial-rubber mixture shall be determined and agreed directly with the consumers — manufacturers of industrial-rubber products.

It is unacceptable to replace the technical carbon by shungite product at the initial stage of implementation because the composition, properties and coarseness of the latter and technical carbon are different. This will be revealed during long operation.

So, the coarseness of the applied shungite rocks and their structural interrelations shall be studied thoroughly and controlled when sending the original raw materials for processing. It may appear that shungite rocks from different deposit sections contain silica of various coarseness. Such rock can't be used in industrial-rubber goods.

Fast correction of the coarseness, mineral and chemical composition of this rock allows improving the stated faults and minimizing the risks of potential negative impact. This is necessary for developing and further stabilization of the technological process.

It has been found that different types of shungite rocks — massive, brecciated and fissile — contain silica of different coarseness. Silica particles have different shapes in all types of shungite rocks. This determines the difference between the technological properties of these rocks depending on the area of application. Every type of shungite rocks shows different technological properties during all-sliming in the counterflow jet-type mill. This has been proved by our researches including differential scanning calorimetry and thermogravimetric analysis.

Different coarseness of silica particles in each rock type has been confirmed by mineralogists from VIMS. Thus, massive rock usually contains silica with coarseness of 1−2 microns, brecciated rock — silica with coarseness of 10−12 microns [Maslakov, et al., 2005]. Also shungite rock contains different polymorphic modifications of silica [Khvorova, Dmitrik, 1972].

Almost all types of shungite rocks contain veined silica with larger flakes. Its physical-chemical characteristics are rather different from those of the finely-dispersed silica in the main matrix of the shungite rock. Surely, in the process of fine grinding, these types of silica look differently and the physical-chemical characteristics of the original also differ. This will influence their further application in different processes. It is especially important when using finely-dispersed shungite rock in industrial-rubber goods.

The dry flow chart of shungite goods production approved by the regulation of VIMS allows controlling only the coarseness properties while structural and qualitative composition of the obtained shungite products using this rated technology can't be corrected. Without correcting the chemical composition it is impossible to get a product with pre-defined properties. So additional testing for improving the quality of the original is required.

All attempts of changing the qualitative composition of the original shungite products with standard enriching technological methods — gravitation, flotation, magnetic separation, radiometric sorting etc. were useless, despite of different technological properties of all mineral components in the main matrix of the shungite rock.

This occurs due to the small coarseness of all components of shungite rocks. This requires a rather fine grinding. Usually this requires a rather expensive and labour-intensive process of further separation of finely-dispersed products. Many institutes have performed great work in this field, in particular, Mekhanobr Institute, KarRC RAS Institute of Geology and VIMS, but there were no stable positive results which can be implemented in the industrial practices.

So, a new task has emerged: to find new ways of improving the technology of shungite goods production with change in their qualitative indicators which depend on the content of the carbonaceous matter, silica, carbonates, silicates, aluminosilicates and sulphides which create an acidic medium during the processing. It influences further application of shungite products.

The most simple way of correcting the quality of shungite products is blending (averaing) of shungite components with different structure and properties. Products with defined chemical composition can be obtained by changing the structure and properties of two mixed rocks. So let us consider the blending technology with the obtaining of potential products with the defined quality.

Mechanical and chemical leaching is the best method for obtaining a high-quality product. The considering of these problems is of great theoretical and practical interest. Such a task is set for shungite rock for the first time.

Below we will consider some variants of the correction of quality indicators considering the changes of the substance structure, chemical composition and potential influence on further technological processes.

3.2. Improving the Quality of Obtained Products by Transformational Blending (Averaging) of Shungite Rocks

The content of carbonaceous matter and silicious connections plays an important role in the technological assessment of shungite rocks. The designed technological manufacturing of shungite products is planned to be applied for shungite rocks with 25–45 % of carbonaceous matter, 45–70 % of silica and 5–10 % of silicates.

As it has been mentioned already, the content of carbonaceous matter is 0.5–98 % in the deposit. The total volume of shungite rocks with 25–45 % of carbonaceous matter is 35 %. Other rocks of the deposit with 0.5–25 % (low-grade) and 45–98 % (high-grade) of carbonaceous matter are not included in the designed composition.

To increase the total volume of shungite rocks which corresponds to the rated data the low-grade and high-grade types shall be blended before the processing. The blending technology shall be applied so after that the required content of the carbonaceous matter and silica in shungite rock shall not exceed the design target values or rules provided by technical conditions. For this, the chemical composition and properties of all components of the shungite rock shall be studied and volume variants of their mixing ahould be chosen before blending. Thus, a small quantity of carbonates or shungite rocks with a large content of sodium-potassium compounds shall be added for reducing the acidic features of shungite products. They will neutralize the shungite rock acidic features.

The volume of mixed products shall be defined experimentally, therefore full analysis of all shungite products for mixing shall be performed. Not only the amount of carbonaceous matter and silica shall be defined but also of other components: carbonates, silicates, aluminosilicates etc. After the blending the obtained mixed product shall also be checked for compliance with the technical conditions and sent for selling..

Blending allows the increase of the total volume of the applied rock on 70–80 %. Remaining other 20 % of shungite rocks contain a small amount of carbonaceous matter (0.5–15 %) but also a lot of carbonates, silicates, aluminosilicates. They can be used in other fields. This amount of shungite rocks can be used as a raw materials for liquid glass, concrete, haydite, foamed concrete, lime and light weight concrete productions.

Shungite rocks with high content of carbonaceous matter (65–98 %) can be used as carbonaceous fertilizers, a pigment for making black paint and for medical purposes. Such rock serves as a very efficient fertilizer for growing potatoes and other vegetables. It is supported by long-term practice [Toikka, 1946].

Low-grade shungite carbonate rock or rock with high content of sodium-potassium compounds also can be used for deoxidating soil. The carbonaceous matter in shungite rock also can be used as an artificial fertilizer.

Blending of shungite rocks can be made at the pit by loading the haulers at the sections of rock development with different content of carbonaceous matter.

The mixture shall be made with such ratio so that the content of the main components of shungite rocks — carbonaceous matter and silica — corresponds to the rated composition. For specifying the mixture parameters, the content of carbonaceous matter and silica of low-grade and high-grade rock shall be previously studied and then the volume of mixing shall be calculated.

The volume of each truck load with different rocks is determined according to the chemical composition. The rocks are mixed efficiently when loading the trucks with a scoop shovel and unloading from the haulers. The shungite rock is taken to the processing enterprise after blending where it undergoes additional chemical analysis. Also the need of further blending to the required conditions is determined there.

Blending of shungite rocks at the pit is made as follows. Firstly, the deposit sections where the shungite rocks significantly differ in the chemical and mineral composition are chosen; then the necessary volume of their mixing shall be defined. Thus, the rocks from different deposit sections will be loaded into one truck. As mineral processing is not provided in our case, the rock for processing shall be a subject for a process control.

If the rocks have rather different chemical composition it is necessary to blend them by mixing separate batches at the reserve area of the raw materials' consolidation centre at the processing enterprise.

Usually, additional blending is made before the rock transferring to the receiving bunker. The level of mixing, i.e. blending, shall be estimated by chemical analysis according to the above-listed technique. Blending can significantly increase the total volume of all rocks (up to 80 %) which will correspond to the technical conditions.

For the blending of shungite rocks with 20 % of carbonaceous matter, it is not recommended to use low-grade shungite rocks which contain less than 15 % of carbonaceous matter but a lot of silicates, aluminosilicates, for the rock main bulk blending because they have many harmful mixtures which negatively influence their further application, especially in rubber production. The rocks which contain less than 5–7 % of carbonaceous matter shall be used for theproducing of other important materials such as liquid glass, concrete, haydite, foamed concrete and light weight concrete.

The qualitative composition of shungite products can be changed not only by blending, but also by mechanical separation or chemical leaching. For this, the technology of converting of separate chemical components into other phases (e.g., carbonates into soluble salts), should be used.

Leaching from the shungite rock can also be used for removing of all sulphides and other sulphur-containing components which negatively influence the quality of the original products. But to apply the leaching process on the first stage, one has to know the procedure of acid generation in the shungite rocks. The knowledge of the physical-chemical basics of acid generation will allow the developing of leaching technology.

The process of sulphides dissolving in the being processed shungite rocks can increase significantly in the case of temperature increase of being processed rocks. This is especially remarkable during the drying process. Due to this, we've found it necessary to consider the issue of sulphide behaviour during the drying process, in particular, their dissolution under different temperature modes. During this process, most sulphides are decomposed as affected by the temperature.

But the process of sulphide dissolution during drying is still unexplored. So before dealing with the issues of leaching, we will briefly consider the procedure of acid generation of shungite rocks and the sulphide behaviour during technological drying, the procedure and the technological characteristics of the processes of mechanical leaching and also the process of carbonates chemical dissolving. The obtained shungite products will have new properties and features which can be used in several fields.

3.3. Procedure of Acid Generation in Shungite Products and their Role in Sulphide Leaching Process

All shungite products have an acidic reaction when obtained by process operations (crushing, grinding and size specification). The pH of these products is usually 3.5–6. The acidic reaction of shungite rock is usually determined by sulphur compositions. The procedure of generation of sulphur compositions in shungite rock is currently unexplored. There is only some information about generation of sulphuric acid during products storing. Due to this, we have set a task for study of their features which appear during processing: crushing, grinding, size specification, drying and storing.

Usually, in shungite goods production which arrive for processing, the moisture of shungite rock is 5–8 % (sometimes — up to 11 % when it is raining). The pH of the shungite rocks decreases significantly, especially if they contain moisture, and is 4–5. Further during crushing, grinding, size specification and drying the pH of the original products is directly proportional to the coarseness of the particles. The acid condition is increasing during storing of the obtained finely-dispersed products. Practice has showed that the original products shall be stored in air-proof containers which protect them from oxygen and moist air for stabilizing the acidic condition.

The influence of the percentage of the carbonaceous matter on pH of shungite products has not been determined so as the role of sulphides in acid generation. These works shall be performed on several products in laboratory conditions and under industrial operations.

From our point of view, acid generation in shungite rock depends on the physical-chemical characteristics of the organic carbonaceous matter and additional mineral inclusions into the shungite rock main matrix (first of all — sulphides).

The chemical composition of organic concentrates has been thoroughly studied in Russian Geological Research Institute, Federal State Budgetary Institution (analyst is M. V. Bogdanova) as presented in Table 3.3.

Table 3.3. **The chemical composition of organic carbon-bearing products**
(according to Russian Geological Research Institute,
Federal State Budgetary Institution)

Product (place of sampling)	Humidity, 105/180 °C	Ash content	Sulphur content		Specific gravity in gramm per cm³	Volatile content, V	Chemical components on anhydrous basis				
			Total	Sulphides			C	H	N	S	O
	%						%				
Massive maksovite (Maksovo)	6.9/7.4	7.5	5.8	5.7	–	4.9	98.36	0.45	0.63	0.62	0.24
Idem	6.8/7.8	7.2	3.4	3.1	1.96	4.6	97.14	0.42	0.72	0.45	1.27
Idem, graphitoid	–/7.2	8.3	3.6	3.2	1.97	4.2	96.53	0.59	0.70	0.68	0.15
Idem (Shunga)	–/1.1	9.6	4.0	3.5	1.89	3.5	97.16	0.59	0.69	0.64	0.92
Shungite (Shunga)	–/1.6	10.2	4.0	2.9	1.90	3.6	96.79	0.54	0.76	1.22	0.69
Idem	–/1.7	5.4	1.9	0.7	1.97	2.6	96.62	0.48	0.71	0.66	1.53
Lyddite (Maksovo)	2.9/3.4	1.2	0.3	0.1	2.12	1.6	98.30	0.34	0.58	0.32	0.44
Silicious tuff (Maksovo)	4.1/5.4	1.1	0.2	0.1	2.08	2.1	98.42	0.33	0.51	0.18	0.58
Shale (Maksovo)	2.8/4.0	4.7	3.6	3.3	1.93	3.2	98.69	0.31	0.51	0.35	0.14
Siltstone (Maksovo)	8.1/8.5	6.9	4.3	4.3	2.0	4.8	96.92	0.31	0.69	0.53	1.55
Tuff siltstone (Maksovo)	2.4/3.1	4.1	11.8	10.8	–	8.3	97.53	0.39	0.52	1.06	0.50
Anthraxolite (Maksovo)	2.4/3.3	1.2	0.5	0.3	1.97	1.9	97.73	0.64	0.78	0.34	0.71
Idem	5.4/6.1	1.1	0.1	0.1	1.91	2.0	97.52	0.66	0.81	0.33	0.68

The shungite substance contains 97–98.5 % of organic carbon, as shown in the table. The test results prove that all samples of the concentrate taken from different deposit sections have approximately equal content of carbonaceous matter.

Shungite substance contains 0.3–0.4 % of hydrogen. Also, all samples of rock concentrates with 98 % of carbonaceous matter contain 0.34–1.55 % of elemental sulphur, but usually sulphur is found in sulphides. The content of oxygen is 0.24–1.55 % on anhydrous basis. The content of nitrogen in carbonaceous matter is more stable and is 0.51–0.81 %.

During burning the organic substance, nitrogen and sulphur deflagrate and evaporate in the form of gases. The content of ash in concentrated samples after burning of organic substances is approx. 7–8 %. The ash volume depends on the presence of a big number of mineral mixtures in shungite rock across the whole area of Zazhogino deposit.

Mineral mixtures contain Na, K, Fe, Al, Cu, Ca etc. As a rule, they are distributed across all carbonaceous matter bulk. Their volume is rather different depending on the location within the deposit — from 1 to 30 %. Besides that, shungite rock contains a lot of trace elements — molybdenum, vanadium, strontium and many others which are also irregularly distributed in the rock. Shungite rock contains such mineral inclusions as roscoelite, karagonite, brassil, capillose, vegodarite, blinde, copper ore etc. sized from 2 to 7 microns. Carbon, hydrogen, nitrogen, sulphur, oxygen can be easily removed from the organic substance during its transformation. It is very difficult to remove less-common and disseminated elements from organic substance probably because they form part of the fullerene carbon structure.

As a more active element sulphur quickly reacts with other substances changing the oxidation-reduction potential of the whole system. The carbon and sulphur interreact easily forming high-reactive carbon disulfide compounds. They remain active even under room temperature and are formed in carbonaceous matter when reacting with oxygen and water leading to an acidic reaction.

The acid condition of shungite rock is positively related to term of storage and inversely — to the coarseness of grinding. Also, it was found that the increase in the acid condition depended on the percentage of the carbonaceous matter and the content of sulphur.

If the shungite rock contains carbonates the above-given regularity does not occur. Vice versa, during storing carbonate shungite rocks their alkalinity is increasing.

All above-listed sulphur chemical modifications are very active and easily participate in all oxidation-reduction reactions occurring during grinding of shungite rocks.

pH of non-carbonate rocks which contain approx. 1.5 % of total sulphur is 5.5. When its content exceeds 3.5 % pH will reduce to 3.5.

We have checked this phenomenon on the shungite non-carbonate rock rich in carbon which contain 62.1 % of carbonaceous matter, 3.4 % of total sulphur, among which sulphide sulphur comprises 2.1 % and silica — 28.6 %.

Three samples of different coarseness classes have been tested; they were taken from different sections of the VSI mill: air classifier (60–70 μm), cyclones (20–25 μm) and sock filter (5–10 μm). The three samples were mixed with a standard agitator. Every 20 minutes we took samples for determining pH. The suspension contained 20 % of solid substance.

Fig. 3.3. shows the changes of the pH solution since mixing the shungite in the distilled water. The obtained data clearly show that the level of dissolution of the sulphur-containing compounds increases rapidly at the initial stage and after 20 minutes of mixing it reduces. It was found that the acid condition was directly proportional to the fineness of grinding. This regularity is peculiar for all samples with different size distribution, but in the samples with more fine grinding the pH value solution is reduced much faster than in coarse-ground samples.

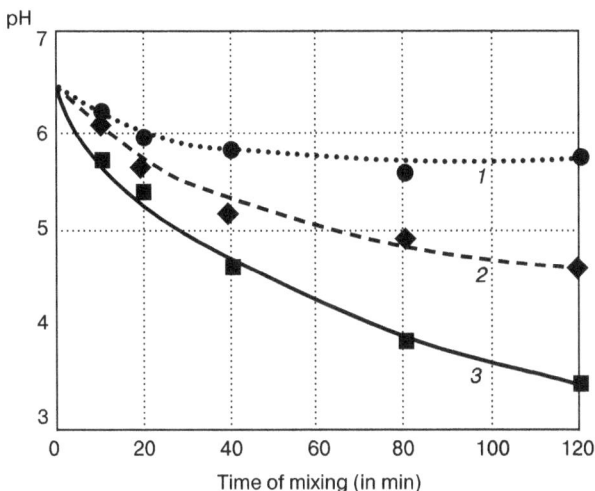

Fig. 3.3. Dependency between the changes of pH values and the period of mixing particles of shungite rock:
1 — air classifier; *2* — cyclone; *3* — sock filter

These tests have proved that the acid features of the solution of shungite rocks depend on the content of sulphur-containing connections and the level of grinding.

Oxidizing of the sulphur and forming H_2SO_4 is very simple and is made according to the following reactions:

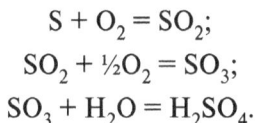

$$S + O_2 = SO_2;$$
$$SO_2 + \frac{1}{2}O_2 = SO_3;$$
$$SO_3 + H_2O = H_2SO_4.$$

When sulphur oxides, not only sulphuric, but also sulphurous acid is formed. So, all carbonaceous shungite rocks have an acidic reaction. The acid condition depends on the content of sulphides and elemental sulphur. The formed sulphuric acid easily reacts with iron sulphides decomposing them to elemental sulphur or sulphur-containing gases. This process can run even under room temperature and pH<4. When pH increases from 4 to 8, brassil will not decompose.

Brassil decomposition with different pH values is shown on Fig. 3.4.

When sulphides oxidize the oxidation-reduction potential (Eh) increases from 0.6 to 1.2 V.

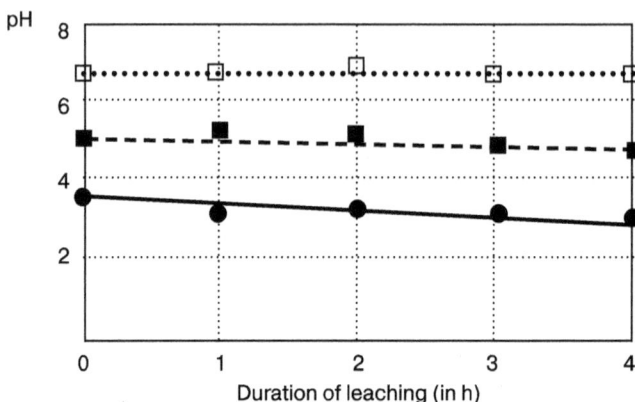

Fig. 3.4. The influence of pH-medium on the kinetics of sulphide leaching

The process of sulphur creation in carbonaceous matter took place on the transformation stage of the putrid ooze when hydrogen sulphide oxidized and the elemental sulphur formed according to the following reaction:

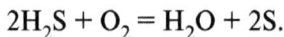

$$2H_2S + O_2 = H_2O + 2S.$$

After that the sulphur reacted with oxygen and sulphate ions were formed.

$$S + O_2 = SO_2.$$

Further sulphate ions in the alkaline medium with 4–8 pH reacted with metal ions forming several sulphides.

According to F. V. Chukhrov [1955], marcasite is crystallized in acid medium with 4–7 pH and brassil — in an alkaline medium with 7–8 pH. As shungite rock includes both brassil and marcasite. the process of crystallization of sulphides took place with 4–8 pH.

The carbonate and sulphate components have also changed during sulphide crystallization. Specifically, the carbonate alkalinity of water (HCO_3) has reduced and the sulphate acid condition increased. During this process both calcium carbonates and sulphides were formed. When changing pH further sulphites and sulphates were oxidized according to the following reactions:

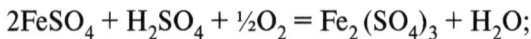

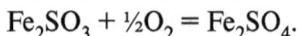

$$2FeSO_4 + H_2SO_4 + \frac{1}{2}O_2 = Fe_2(SO_4)_3 + H_2O;$$
$$Fe_2SO_3 + \frac{1}{2}O_2 = Fe_2SO_4.$$

Marcial water near Karelian shungite deposits contains a lot of sulphates and elemental sulphur, the water of the spring is oxygenized.

Tests on oxidizing and dissolving brassil in the oxygen-containing water medium were conducted. After shaking 100 g of brassil in a 0.5 l flask for one hour under normal conditions, 0.54 mg-equ/l of SO_4 ion was found, specifically the pH of the solution fell from 6.5 to 4.2. The sulphuric acid formed by oxidation of elemental sulphur caused the brassil solvation the.

When a similar mixture was heated to 80 °C and shaken again the pH value reduced to 3.6 %. These tests showed that the solubility of brassil had increased in hot water. The ferrous sulphate which formed by the dissolution process catalyzes the decomposition of brassil reacting with it. This can be the best observed during shungite products storing. When shungite rock is stored in humid premises rust-ochroid iron oxyhydroxides are formed.

So, on the basis of the considered material we can conclude that the elemental sulphur in the main carbonaceous matter of the shungite rock can be found in all samples and can easily react in chemical interreaction with the oxygen in the air, forming sulphate-ions and sulphuric acid. These reactions proceed rather quickly under normal temperature.

Sulphides in shungite rocks react with the sulphuric acid and educe elemental sulphur. Sulphides are oxidized under room temperature of the ambient air (10−25 °C) according to the following reactions:

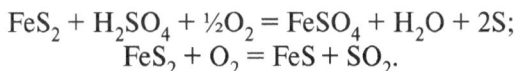

$$FeS_2 + H_2SO_4 + \tfrac{1}{2}O_2 = FeSO_4 + H_2O + 2S;$$
$$FeS_2 + O_2 = FeS + SO_2.$$

Other sulphides also oxidize in the air, but this process is very slow. For catalyzing it one has to make the temperature higher.

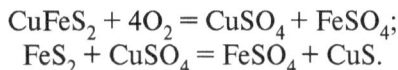

$$CuFeS_2 + 4O_2 = CuSO_4 + FeSO_4;$$
$$FeS_2 + CuSO_4 = FeSO_4 + CuS.$$

Copper as a valuable element can be recovered from the solution with the help of broken iron.

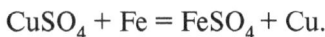

$$CuSO_4 + Fe = FeSO_4 + Cu.$$

The main peculiarities of physical-chemical characteristics of oxidation, passivation, stablization and decomposition of sulphides as applied to the processes of flotation have been articulated by V. M. Avdokhin, Professor of MSMU [Avdokhin, Abramov, 1989].

His paper deals with issues of activation and passivation of the surface of sulphides of various origin. It also considers the impact of one, two and more oxidizers and ways of improving the flotation processes when mineral processing sulphide ore; it deals with the main thermodynamic characteristics of the reactions of sulphides oxidizing as applied to flotational mineral processing.

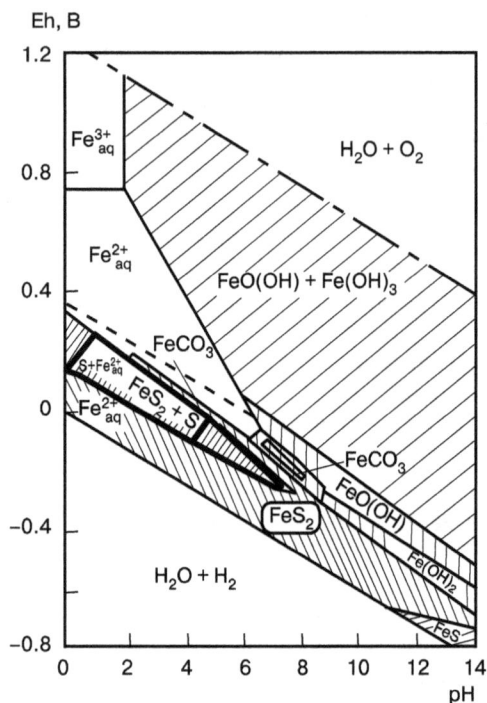

Fig. 3.5. Chart of thermodynamic stability of iron sulphides in aqueous solutions under 25 °C and 10^5 Pa, according to [Avdokhin, Abramov, 1989].

As for leaching of finely-dispersed sulphides inside the tiny fractures of shungite rock, the main physical-chemical and thermodynamic provisions developed for flotation by the authors of the above-given paper are similar and can be applied also for explaining the leaching processes. This concerns the chart of thermodynamic stability of iron sulphides in aqueous solutions under 25 °C and 10^5 Pa and the chart of activity of the dissolved sulphur equal to 10^{-1} (Fig. 3.5).

[Krauskopf, 1956] paper deals with the processes of iron oxides and sulphides forming in water under the temperature of 25 °C, pressure of 10^5 Pa and the activity of the dissolved sulphur equal to 10^{-6} [Krauskopf, 1956]. The chart of the relation between oxides and iron sulphide stability in water according to this data is represented by Fig. 3.6.

As we can see from the chart, brassil is formed and can exist only in a reducing medium with pH from 4 to 9 and Eh from 0 to 0.4. If pH is less than 3.8 brassil is not formed, but decomposed. Also, brassil is not formed and can not exist in the reducing medium with pH exceeding 9.

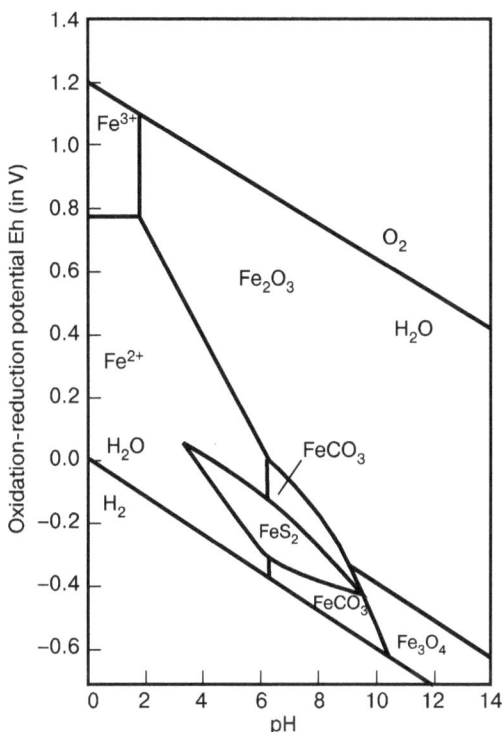

Fig. 3.6. Chart of Eh–pH ratio for ferriferous products according to [Krauskopf, 1956]

Iron oxyhydroxides can form not only in a oxidizing medium, but also in the reducing one if pH increases from 0 to 6.

As we can see from the chart, brassil can exist only in a reducing medium with pH from 4 to 9 and Eh from 0 to 0.4. If pH is less than 3.5 brassil is decomposed educing sulphur. Also, brassil is decomposed and can not exist in the reducing medium with pH exceeding 9.

The considered chart proves that brassil decomposes quickly in strong acidic and alkaline media forming sulphates or reduces to elemental sulphur. This increases the acid condition of ferriferous components of shungite products.

Thus, on the basis of the obtained data we can conclude that acid in shungite rock is generated due to complex physical-chemical interactioninteractions between sulphides. elemental sulphur and air oxygen. The speed of acid features forming depends on the air humidity. So shungite products have to be packed and stored in airproof containers.

3.4. Technological Characteristics of Leaching of Sulfides and Other Components from Shungite Rocks

Practice showed that during crushing, grinding and specification of shungite rocks, especially if they contact with air oxygen, they obtain the acidic features. During further application these acidic features have certain negative impact and sometimes narrow the scope of potential use of shungite products.

In case of very fine grinding of shungite products, gravitational, flotational, magnetic and radiometric methods of their separation have low economic and technological efficiency.

For removing sulphur and sulphides from the main matrix of shungite rock, mechanical and chemical methods will be the most practical ones.

The easiest way of reducing the acid condition of shungite rocks is mechanical removal of sulphur-containing compounds. This is done as follows: ground original shungite rock is placed into the conditioning tank where the suspension is mixed intensely with an impeller so the acid-containing sulphides are washed out. The pH of the suspension reduces from 6 to 3, so sulphides become more soluble. 3—4 pH is the best value for dissolving sulphide-bearing minerals. Further reducing of pH values is economically unviable.

If pH is 4—8.5 brassil will not dissolve, but, vice versa, new crystals will be formed from the acidic sulphur-containing solution. But in the process of active mixing of ground shungite powder the brassil is dissolved if pH is 8—14.

If the brassil content in the shungite rock increases the leaching becomes more efficient. Brassil dissolves in the rock if its content reaches 0.5 %. For catalyzing the dissolution of the rock bare in brassil, firstly, the solvent shall be additionally acidated. Ferrous sulphate can be used for acidation to reduce pH of the solvent to the optimal value of 3—3.8.

The offered technology can be easily implemented in the industrial practice. Specifically, the leaching process can be catalyzed by adding special chemical additives such as chlorohydric or sulphuric acid, ferrous sulphate etc.

Oxygen shall be added for activating the process of dissolving the sulphides. This can be done by blowing off pressurized air through the suspension processed. By interreacting with the suspension the air oxygen will accelerate the process of dissolving the sulphides and the leaching will be significantly catalysed.

It is recommended to use sulphuric acid or ferrous sulphate as catalysers for leaching sulphides from shungite rock. According to the Eh–pH ratio chart for ferriferous products [Krauskopf, 1956] these components shall be included only if pH is 3–3.8.

Usually the product with the set range of pH from 3 to 3.8 can be obtained without adding sulphuric acid or ferrous sulphate by intense mixing of the suspension of the shungite products at the ratio L:S= 1:1 in the conditioning tank. In this case for catalyzing the process of leaching the sulphides air is blown into the conditioning tank so that the oxygen accelerates the dissolution process.

Chemical additives (sulphuric acid or ferrous sulphate) shall be included into the suspension of the shungite product only if the original rock contains less than 1 % of sulphides. This occurs because the content of sulphides in shungite rock changes directly proportionally to their leaching and pH increases to 4 only. For supporting the normal progress of leaching the value of pH shall be approximately 3–3.8. So, for reaching the best pH value ferrous sulphate shall be added to the original suspension.

Economics is the key aspect of this process, i.e. the cost of solvents and potential losses; the process should be simple and the quality of the products should be high, i.e. minimal hydrogenization should be reached.

The sulphur-containing products dissolved during hydrogenization shall be separated from the solid phase of shungite rocks. The liquid phase obtained by leaching shall be neutralized and the sulphides shall precipitate out. Calcium and magnesium carbonate can be used as precipitators for acidic ferriferous solutions. The sedimentation proceeds according to the following reactions:

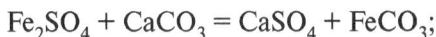

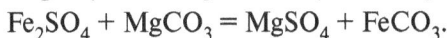

$$Fe_2SO_4 + CaCO_3 = CaSO_4 + FeCO_3;$$
$$Fe_2SO_4 + MgCO_3 = MgSO_4 + FeCO_3.$$

Precipitates such as $CaSO_4 \cdot 2H_2O$ or $MgSO_4 \cdot 2H_2O$ are removed by filtering and ferric carbonate as affected by ferrous sulphate is decomposed educing carbon dioxide.

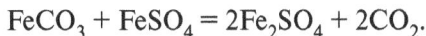

$$FeCO_3 + FeSO_4 = 2Fe_2SO_4 + 2CO_2.$$

The process of dissolving sulphides by mixing is inconvertible. This allows removing of almost all sulphide minerals from the rock by mechanical methods in the acid medium. This method of leaching has been offered by us for the first time. We consider that it shall be implemented into industrial practice of shungite rock processing for improving the quality of the obtained shungite product.

Besides sulphides, in some cases other components of shungite rock should be also removed (e.g., carbonates) which form complex compounds when dissolved.

The process of complex compounds forming during dissolution of shungite rocks is rather diverse and interesting from the chemical point of view. The dissolution process itself is smooth and inconvertible. After this. shungite products without acidic features are obtained after leaching.

It is obvious that study of the solubility of sulphides in the shungite rock is of great theoretical and practical interest and is required for commissioning and operation of pilot production.

Carbonates can also be removed from shungite rocks. It is well-known that the solubility of calcium carbonate in sodium sulphide solution with concentration of 0.05 mole/l increases by 320–330 times. If the concentration is increased to 0.09 mole/l the solubility increases by 2330 times. This method can be used for removing of carbonates from shungite rocks. Such technology of processing can appear very prospective and it shall be checked in industrial conditions.

The solubility of sulphides contained in shungite rocks will increase by many times over in case of using not pure water, but an oxygenized electrolyte solution. In this case, the partial solubility of shungite rocks may increase by several times. However, these works have not been industrially confirmed yet.

All works on mineral solubility contained in shungite rock were performed only in laboratory conditions with the process qualitative assessment. There is not many industrial data with qualitative assessment. These works shall be extended and performed together with future consumers of shungite products.

Currently there are many works on solubility of mineral salts (in particular, in salt working) present in low- grade shungite rocks. But the questions of leaching of soluble salts have not been raised yet, although they are also worth to study.

To conclude, we will repeat that all sulphur-containing compounds can be dissolved almost completely by mixing the suspension of the acid liquor of shungite rocks which contains 1–3 % of sulphides and elemental sulphur in the conditioning tank. Then the above-listed compounds can be removed by separating of the liquid phase and further sedimentation. The level of leaching in the conditioning tank is directly proportional to the suspension pH. According to the Krauskopf's chart of solubility during leaching the best concentration of pH shall be not be less than 3.8.

To prove the above-listed, here is an example for solubility of shungite rocks with 31.5 % of carbonaceous matter, 55 % of silica and 1.8 % of sulphur-containing products. In case of a 15-minute contact of a fine shungite rock with air the content of total sulphur has reduced to 1 % and concentration of SO_4 in the solution has increased to 350 mg/l, while the pH value of the suspension has reduced to 3.5.

Analysis of physical-chemical characteristics of shungite rock during sulphide dissolution has shown that these works are rather prospective. Abroad, such works are already used in the industrial practice at many coal industry enterprises. Due to this, we consider that we have to analyze the results of these works.

3.5. Opportunity to Apply Established Technologies for Removing Sulphides from Coal Rocks

The process of removing sulphur from sulphur-containing coal has been used in the industrial practice for a long time already in leading countries of the world: USA, Japan, Germany, etc. [Yagodkina, 1984]. Meyer technology (USA) has been described in literature. The bottom line here is that sulphur sulphides can be leached from coal bferric sulphate ($Fe_2(SO_4)_3$) aqueous solutions under temperature 100–130 °C.

Using this method approximately 95 % of sulphides can be removed from coal with 3–4 % of ferric sulphides. Only 47 % of other components (e.g., total sulphur) is washed out. So harmful mixtures of other sulphides such as copper, lead, arsenic, cadmium can be also removed. So, the Meyers process can be recommended for removing all sulphur sulphides from shungite rock.

Lately the technology has been significantly improved. Processed finely-dispersed coal with sulphides has been processed additionally in anthracene oil under temperature 425 °C and pressure 14 MPa for 1 hour. As a result of such processing coal concentrate with 0.85 % of total sulphur has been obtained from coal with 3–4 % of total sulphur. Specifically, total recovery of sulphur-containing products from coal rocks using this technology reached 90–95 %.

There are also other methods of hydrothermal acid processing of coal for reducing the amount of sulphide sulphur. In this sphere Battel technology (USA) should be noted. According to it the finely-dispersed coal with 3–5 % of sulphur is mixed with alkaline aqueous solutions. Sodium hydrate and caustic lime are used as alkali.

Coal suspension processed by this technology is heated in a steam chamber additionally under temperature 220–350 °C and pressure of

2.5–17.5 MPa for 30 minutes. Further after processing in the steam chamber the suspension is transferred to the centrifuge. Using such a technology a pure coal product with approximately 0.35–0.65 % of sulphur can be obtained from coal with 3–4 % of sulphide sulphur while at least 90 % of sulphur is being recovered. Other sodium and calcium aqueous solutions can be used instead of NaOH and KOH alkali.

Leaching of sulphides from sulphur-containing coals can be made both in acidic and alkali medium. This has been confirmed by the above-listed theoretical data. The considered physical-chemical characteristics of leaching of sulphur from sulphur-containing or shungite rocks fully comply with the Krauskopf's chart of solubility.

The above-listed technologies can be used for removing of not less then 90 % of all sulphur including sulphide one from shungite rock of Zazhogino deposit.

Besides the above-listed chemical components also other chemical compounds can be used for leaching sulphur, e.g., sulphuric, chloro-hydric, nitric acid, calcium and ferrous chlorides, ammonium and so-dium carbonates, etc. As it was mentioned earlier, the best pH value of acid liquors for leaching sulphides from shungite rocks shall be at least 3–3.8 and more than 8 for alkaline solutions. Under such conditions other compounds can be used for sulphides dissolving (e.g., peroxy and aqueous solutions of ferrous sulphates).

All potential components for desulphurating of shungite rocks should undergo industrial testing.

During leaching of sulphides, presence of the dissolved gas in the liquid phase of the leached suspension plays a big role. The kinetics of the dissolution process depends on the composition and characteristic of these gases. Due to this we will consider the influence of the dis-solved gases in the suspension on the sulphide leaching process.

3.6. Role of Dissolved Gases in Process of Sulphide Leaching from Shungite Rock

The main component for leaching of sulphur-containing miner-als and sulphides from shungite rock is water. Moreover, the chemical composition of such water and its dissolved gases influences the process significantly.

Usually the water used for leaching has a complex chemical content with different concentration of hydrogen ions which influence the com-plex process of oxidation-reduction reactions. The original water contains a lot of dissolved gases — O_2, N, CO, CO_2, HCO_2, H_2S, S, SO_2, SO_3, which can influence the leaching process both positively and negatively. Their composition and number change depending on the water pH.

Usually water consists of carbon-dioxide compounds, the content of which strongly depends on the pH, i.e. carbonic acid can be represented as gas CO_2, non-dissociated molecules of H_2CO_3 or as HCO_3^-, and CO_3^{2-}. All these forms of carbonic acid are dynamically balanced depending on pH which can be represented in form of a reaction:

$$CO_2 + H_2O \leftrightarrow H_2CO_3 \leftrightarrow H^+ + HCO_3^- \leftrightarrow 2H^+ + CO_3^{2-}.$$

These relations can be represented as a chart (Fig. 3.7).

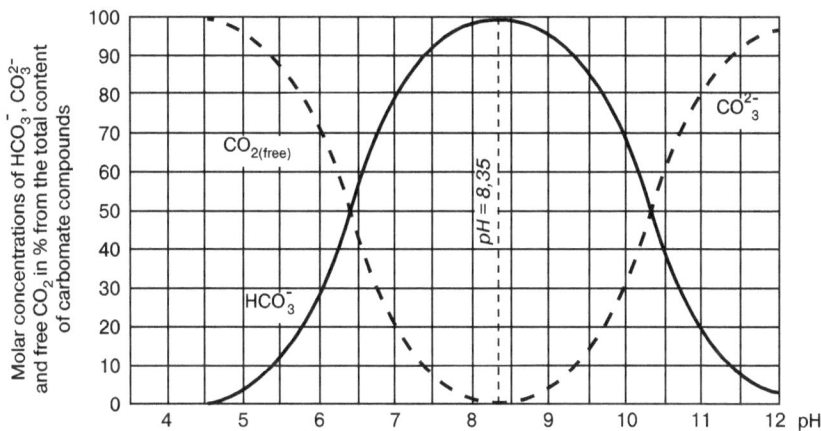

Fig. 3.7. Ratio of molar concentrations of different forms of carbonic acid depending on the pH values of the medium [Kastalsky, 1957].

From the given chart we can see that carbonic acid is educed when pH is at least 4.5. In such case, carbonic acid is in the form of non-associated gas CO_2 which does not affect leaching in any way. Before leaching, carbonates can be easily removed from the solution by decarbonization. An ordinary Pachuca tank can be used as a decarbonizer which is widely used in hydrometallurgy of uranium and rare metals.

So, carbon dioxide is removed by decarbonizing from the suspension within the leaching process with pH at least 4.5; the solution is oxygenated by blowing out the suspension. The oxygenated suspension actively reacts with the sulphides of shungite rock dissolving them. Besides oxygen, the suspension also contains an active gas called hydrogen sulphide which also influences the process of dissolving sulphides and other sulphur-containing compounds. The efficiency of dissolving of sulphides depends on the level of oxygenation and sulphitation of the suspension so as on the medium temperature and speed of mixing the suspension in the conditioning tank. The pH plays a drastic role in dissolving the components; its value shall be approximately 3.2–4.

Dissolved oxygen plays the greatest role during leaching. It reacts not only with sulphides, but also with carbon which is the main component of the shungite matrix. It forms carbon bisulphide which strongly acidifies sulphides and other sulphur-containing compounds.

The water which is received for leaching usually contains oxygen from the air. The water can be oxygenated by blowing it out by air considering that the level of oxygen solubility is rather high in water (Table 3.4).

Table. 3.4. **Oxygen solubility in water during the air contact and atmosphere pressure** [Kastalsky, 1957]

$T, °C$	$O_2, mgpl$	$T, °C$	$O_2, mgpl$	$T, °C$	$O_2, mgpl$
0	14.6	11	11	30	7.5
1	14.2	12	10.8	35	7
2	13.8	13	10.5	40	6.5
3	13.4	14	10.3	45	6
4	13.1	15	10.1	50	5.6
5	12.8	16	9.9	60	4.8
6	12.4	17	9.7	70	3.9
7	12.1	18	9.5	80	2.9
8	11.8	19	9.3	90	1.6
9	11.6	20	9.1	100	0
10	11.3	25	8.3		

The oxygen solubility in water is inversely proportional to the temperature. Under ambient temperature 1–25 °C the oxygen content in water is 8–14 mgpl. This is enough for acidifying sulphides and dissolving them.

Besides oxygen, hydrogen sulphide, bisulfide ion HS^- and silphude ion S^{2-} play an active role in leaching. The ratio of the reduced sulphur forms usually depends on the pH medium (Fig. 3.8). If pH is less than 7, H_2S is prevalent in the solution; if pH increases to 13.8 — ions of HS^- are prevalent and if pH is 13.8 to 14 — ions of S^{2-} [Staschuk, 1968].

As one can see when sulphides are leached from shungite rock with pH from 3.5 to 4.5, sulphur dioxide will be represented as H_2S. This form is rather active and during leaching of shungite rock is an efficient solvent of sulphur-containing compounds. Other reduced forms of sulphur (HS^- и S^{2-}) are less active in the process of dissolving of sulphides.

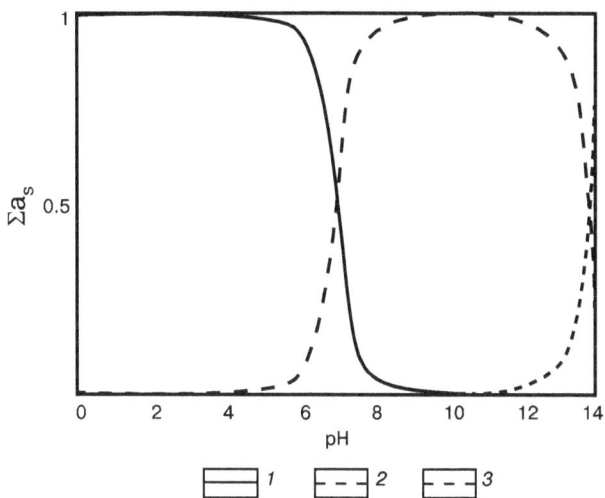

Fig. 3.8. Reduction of sulphur form depending on pH
1 — H_2S; 2 — HS^-; 3 — S^{2-} [Staschuk, 1968]

To conclude it should be mentioned that the active oxygen and sulphur dioxide contained in the leached suspension strongly influence the sulphide solubility and their further potential removal from the leached shungite rock. The air in the mixed suspension is a necessary part of the flow chart of sulphide leaching. So, it is recommended to implement the flow chart of sulphide leaching together with Pachuca tanks used for oxygenating.

3.7. Flow Chart of Sulphide Leaching from Shungite Rocks and its Mechanization

Long-term industrial practice of the largest companies of Western countries (USA, Germany, England, Japan) has proved that leaching can be used for removing sulphides and other sulphur-containing minerals from sulphur bituminous coal.

Also, sulphides can be leached from shungite rocks. During this process pH for acidic media shall be 3–3.8 and for alkaline ones — 8–10. This has been substantiated for shungite rock above. Leaching technology and its equipment are almost the same both for acidic and alkaline media.

For catalyzing the leaching process, the original suspension of shungite rock should be heated to 70–80 °C. For intensifying the process of dissolution of sulphides and elemental sulphur the suspension shall be mixed intensely with additional supply of fresh oxygenated air. Usually the average content of oxygen in the air atmosphere is 20–21 %.

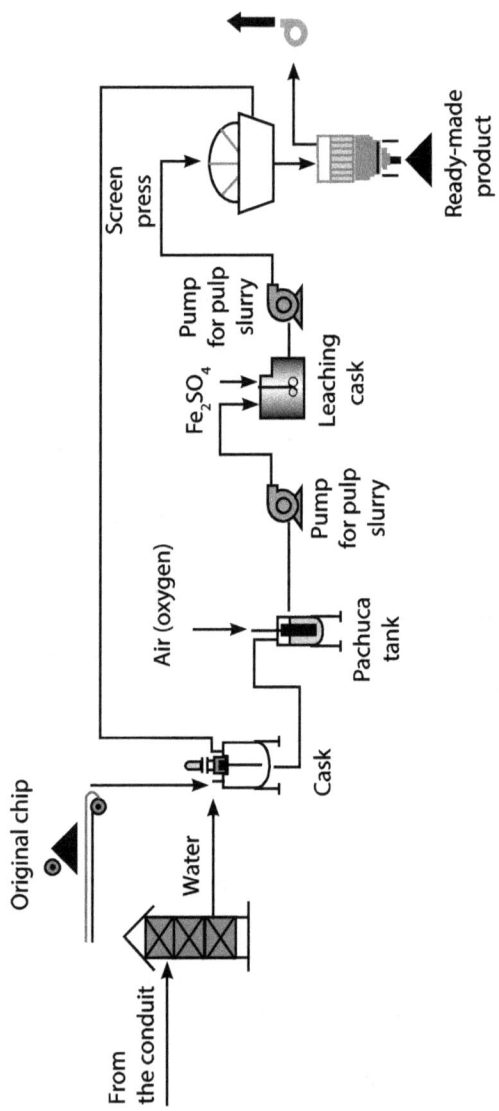

Fig. 3.9. The diagram of the technology of leaching sulphides from shungite rock

During mixing and additional air supply the sulphides and other sulphur-containing minerals are acidified intensely. This can be done in widely-used mechanisms called Pachuca tanks. The structure of the air treatment plant is simple: a tube is inserted into a round cylindrical vessel. The former contains another tube which is by 2/3 smaller than the first one. Pressurized air is supplied through the internal tube. It passes through the narrow tube into a longer and wider one. The air in the wide tube bursts into smaller bubbles, oxygenates the whole volume of the mixed suspension. Apart from air supply reagents for catalyzing the leaching process can be added to the next mechanism — the conditioning tank.

For the final solution of mechanization and the leaching technology industrial tests shall be made for specifying the technological parameters and the time period for processing rocks so as determining the need for additional supply of reagents (e.g., ferrous sulphates). On evidence of industrial tests,thekinetics of the process of sulphides dissolving can be specified and the number of mechanisms necessary for full dissolution can be determined. Refer to Fig. 3.9. for the flow chart of leaching sulphides and its mechanization.

The obtained acid solution will be separated from the main bulk of the shungite suspension by the leaching; after that it will be removed with filters. After filtering the solid phase of shungite rock represented by circulation load again shall be a subject to additional leaching until the desulsphurized product is obtained.

For catalyzing the leaching process, ferrous sulphate is supplied to the original suspension. If there is a big content of brassil in shungite rock, leaching can be made without adding ferrous sulphate. After dissolving f the main bulk of sulphides the acid solution shall be separated by filtering it.

The filtered acid solution is processed by the precipitation agent for removing the sulphur. The formed sulphur presipitate is passed to the tailings. The rectified solution is again returned for leaching. So, sulphides are dissolved and the sulphur residue is removed continuously.

3.8 Peculiarities of the Flow Chart and Mode of Preparing of Medical Shungite Baths

The technology of sulphides leaching from shungite rock is also used for medical procedures implying shungite baths. The sulphur compounds dissolved in water are very efficient for medical purposes.

It is well known that shungite rock is generally deemed to be a material which in certain cases positively influences the human body. Practice has shown that shungites as aqueous solutions, infusions, ointments or pastes were used for treating a wide range of human diseases: skin, nervous, visceral and circulatory diseases [Doronina, 2004; Orlov, 2004; Rysyev, 2004].

This list of diseases proves the wide positive impact of shungite rock on the human body, although its mechanism is still unclear.

The big information flow in the literature written by various experts — physicists, chemists, mineralogists, biologists, etc. — shows that the accumulated practical material in different spheres of knowledge including biological ones sometimes has not been aligned even terminologically. In some cases, it is impossible to define which type of shungite rocks has been used in field researches. This is unacceptable, especially in medicine as the effect of positive impact of shungite rock on the human body shall be specified only with the known chemical and physical properties.

Generally speaking about their peculiarities and intended use, the considered shungite substance has no comparable counterparts, Shungite cures, purifies, revitalizes and protects the human body. It even intensifies plant growth. This is proved by the fact that large potatoes grow at the deposit [Kalinin 1975; Nartov, 1798; Orlov, 2004, Pekki, 1975].

Sulphur solutions have a positive impact on the human body, in particular, on the skin and nerve system. So, sulphur has been used as a drug since ancient times. Until the present it is widely used as a drug for preparing various sulphur-containing ointments and applied for treating internal diseases. Sulphur deficit makes us feeling unwell. Sulphur is contained in protein and various amino acids so as in all muscular cells of internal organs, especially in hair and nails. Sulphur is the main chemical element of bioactive body structures.

It is considered that the impact of shungite rock on the human body depends on presence of fullerenes in the carbonaceous matter [Galdobina, Gorlov, 1975; Doronina, 2004; Yeletsky, Smirnov, 1995; Orlov, 2004]. But the mechanism of creation of fullerens and the peculiarities of their impact on the human body are still underexplored.

Sulphur is the component of all bioactive substances of the human body. It is present in vitamins, hormones, insulin and other compounds. Its deficit leads to bone fragility and psilosis.

Sulphur dioxide is often used for desinfecting and desinsecting.

Sulphur ointments are usually used for treating inflammatory process on the skin and internal diseases.

The above-mentioned ways of treatment by using the suspension of shungite products have been applied in practical medicine for 200 years. When sulphur-containing components are dissolved sulphurous anhydride SO_2 is formed. After further acidifying it is transformed into sulphuric anhydride SO_3.

$$SO_2 + \frac{1}{2}O_2 = SO_3$$

This reaction is significantly accelerated by catalysers. Iron, vanadium, chrome catalyze this reaction. They are present in all types of shungite rocks.

Sulphurous anhydride is easily dissolved in water medium under 20 °C. One volume of water can dissolve approx. 40 volumes of sulphurous anhydride forming sulphurous acid.

$$SO_2 + H_2O = H_2SO_3$$

When sulphurous acid H_2SO_3 is heated the reaction is displaced to the left and the solution educes sulphur dioxide SO_2 again. The weak solution of sulphurous acid serves as an efficient detergent for killing germs. It is used for treating damp cellars, basements, wine casks and fermenters so as for killing mold fungi.

A weak solution of ferrous sulphate $Fe_2SO_4 \cdot 5H_2O$ or blue copperas $Cu_2SO_4 \cdot 5H_2O$ causes a significant impact on the human body. Currently such solutions (1–5 %) are widely used in agriculture for sprinkling trees or shrubs and treating the seeds of graminaceous plants before sowing. This is made for killing harmful fungous spores in the seed grain.

So, on the basis of the provided data it has been determined that shungite rock influences the human body due to sulphurous compounds which help treating of several diseases.

It is recommended to arrange a room for taking acid baths at the industrial plant after its commissioning. For obtaining an acid solution for the bathes a suspension of shungite products with 30–40 % of solid content shall be prepared. The size of the particles of the ground shungite rock shall be less than 50 μm. Ground shungite rock can be mixed in a standard mixer until a solution with pH 4.5 is formed. It is not recommended to apply more acidic solutions with pH less than 4.5 without consulting a doctor.

The solid shungite material is filtered from the acidic suspension formed by mixing. After that it is dried and used as a main product; the liquid phase is used in bathes for manipulation treatment.

Sulphurous water is widely used at Caucasian health resorts — Pyatigorsk, Matsesta, Staraya Russa (Novgorod region) etc. When acid sulphur-containing waters are used in bathes at these resorts an odour of hydrogen sulphide appears; during their standage a dark sludge appears which means the settlement of elemental sulphur. A similar process takes place when shungite rocks are used for bathes. Besides sulphur and sulphides shungite rock contains traces of sodium sulphide Na_2S and potassium sulphide K_2S. Those are strong active chemical reagents with high oxidation-reduction characteristics so, for sure, they cause a great impact on the human body.

So, on the basis of the above-given data we consider that the main medical impact of shungite rocks lies in the influence of several dissolved sulphurous compounds with high oxidation-reduction characteristics.

Conclusions

1. Analysis of physical-chemical characteristics of shungite rocks has proved that several admixtures such as sulphides, carbonates and other mineral inclusions negatively influences the processes. These admixtures strongly narrow the scope of application of shungites so sulphides and carbonates shall be removed from the obtained products, especially if they are used as a filler in industrial-rubber goods.
2. It is found that sulphides and carbonates can be removed from shungite products only by chemical leaching. This simple method removes up to 90 % of all sulphur-containing components from shungite products which contain 2–4 % of sulphides.
   By treating of shungite rocks with chlorohydric acid 95 % of carbonates can be removed. The products obtained after leaching with chlorohydric acid shall not cause negative impact on their usage as a filler in industrial-rubber goods.
3. The kinetics of forming sulphurous compounds in shungite rock has been determined within the performed researches. It is shown that the acid condition of shungite rocks depends on two factors: content of elemental sulphur in the carbonaceous matrix and presence of sulphides in the shungite rock. Elemental sulphur is present in the main matrix of shungite rocks both in solid and gaseous form. In case of the slightest changes of temperature conditions and humidity it actively reacts forming carbon bisulfate and other sulphurous compounds.
4. The mechanism of sulphides dissolving of shungite rocks has been considered and the possible area of their existence with different pH values has been determined based on the Krauskopf's chart of solubility. This shall be strictly considered within industrial operation of shungite rocks and the technology of their leaching.
5. It is shown that sulphide and carbonate leaching shall be performed separately. Chlorohydric acid shall be applied for carbonates.leaching When treating carbonate shungite rocks with chlorohydric, acid calcium and magnesium carbonate dissolution occurs. This leads to forming of calcium or magnesium chloride in the suspension. A saturated solution of calcium chloride shall be separated from the suspension and sent to the atomizing drier where crystals of calcium or magnesium chlorides are formed which can be used as new chemical products.

6. It is shown that oxygen and hydrogen sulphide play a significant role in the process of dissolving of sulphides of shungite rocks. Both gases get into the liquid phase from the ambient air by blowing out the suspension with a barbotage. Carbon dioxide which does not influence the process of dissolving of sulphides, but partly reacts with oxygen, is fully removed from the suspension by blowing out (i.e., the suspension is decarbonized by blowing out).
7. The performed tests allowed developing a flow chart of leaching of the sulphides and carbonates from shungite rocks and make a mechanization for the leaching process.
8. During physical-chemical researches the mechanism of impact of shungite bathes on the human body was substantiated. It is shown that the main medical components are sulphur-containing compounds including sulphides and elemental sulphur which can be easily washed out from shungite rocks and the formed acidic liquid influences the human body during taking the bath. This process is practically similar to well-known hydrogen sulphide bathes widely used in practical spa medicine. We consider that the main health benefit of shungite rocks is related to sulphur-containing minerals which are present in all forms of such rock.

Chapter 4
MANUFACTURING PROCESS ANALYSIS AND DEVELOPMENT OF NEW DIRECTIONS OF SHUNGITE ROCK INDUSTRIAL APPLICATION

4.1. Flow Chart and Its Mechanization

Analysis of researches and the experience of applying shungite products has proven that on the first stage of implementation shungite rock can be used without prior mineral processing. The best chemical composition of shungite rocks which can be used for producing of shungite products without prior mineral processing shall be as follows: carbonacous matter — 25–45 %, silica — 55–60 %, aluminosilicates — 10 %.

In the middle of the deposit the rock volume with the best chemical composition is rather high — at least 50 %. The simplest flow chart has been adopted for their processing: crushing, grinding, size specification and drying. For obtaining crushed products with the necessary chemical content rocks with different deposit sections have been used.

Detailed flow charts of crushing and size specification according to coarseness are widely used at various enterprises in the coal industry. This process has been fully developed and can be applied for shungite rock processing.

The same flow chart with obtaining of the products with different coarseness can be used not only for rocks with the best chemical composition, but also for low- and high-grade shungite rocks with the content of carbonaceous matter from 0.5 to 25 and from 45 to 98 %.

As the production scheme provides obtaining of products with a wide range of coarseness two stages of crushing, all-sliming, size specification and drying have been covered during the development process. Considering that the ready-made products will be of different coarseness two types of bagger equipment are used: for coarse-graded and finely-dispersed products.

The physical-chemical composition of the original raw materials was not considered while developing this flow chart of manufacturing of shungite products as the processes of crushing, grinding and drying of the products with different composition are practically the same.

On the basis of the performed analysis of all previous industrial tests it has been determined that each field requires materials with certain coarseness and chemical composition. Thus, e.g., for foundry-iron, blast-furnace ferro-alloys the coarseness of 10–100 mm is required; for non-ferrous electrometallurgy — 10–40 mm.

For shungite products used as fillers for elastomers in industrial-rubber and tyre industry so as thermoplastics and thermosets, the size of particles of shungite products shall be 5–90 µm; for artificial fertilizers a finely-dispersed shungite product with coarseness from 100 to 140 µm is required; the coarseness of chip-adsorbing agent can vary from 0.5 to 3–5 mm. Shungite products with particles of 1–5 mm coarseness are used in construction for manufacturing radio-shield bricks so as for masonry and plastery.

We deem appropriate to begin implementing of shungite products at velocity filters of main treatment facilities at Kuryanovskaya and Lyuberetskaya aeration station in Moscow replacing arenaceous quartz with shungite chip. For velocity filters of Moscow, shungite chip with coarseness 1–5 mm shall be used.

The shungite product improves the sanitary and hygienic properties of water by reacting with it. Besides that, possessing catalytical and ion-exchange characteristics, shungite product acts as an adsorbing agent reducing the content of phenols, guaiacols, eugenic and oleinic acid in the treated water [Kalinin, Pekki, 1977].

As one can see the range of coarseness of shungite products in usage is very high. So, the designed flow chart considers the requirements of every consumer only in the aspect of shungite products coarseness while at the first stage of the flow chart implementing the material composition of these products is not regulated or corrected.

As a certain product coarseness is required for each certain case, the technology of shungite rocks processing shall be developed considering all potential consumers. It is important for the enterprise to manufacture various products: from large chips to fine powder with a certain chemical composition.

Such technology of shungite products obtaining includes the following operations: two-stage crushing, drying, size specification of the crushed product, packaging the obtained chip. Further all-sliming of the 0.5 mm fractures in a counterflow jet-type mill is possible.

The crushed product shall be classified with a vibratory sieve for obtaining of different fractures of shungite products. The product with particles of coarseness approximately 30 mm is obtained on the first stage of crushing. Shungite chip with the following size distribution is obtained on the second stage of crushing in the roll breaker: 4 mm — 10 %; 0.5–4 mm — 60 %; 0.5 mm — 30 %. This material shall be dispersed through sieves for obtaining of products which the consumers require. From the sieve classifying the product with coarseness 0.5–4 mm is again passed for crushing and particles smaller than 0.5 mm are passed for jet grinding. The crushed material is dried before size specification.

After size specification and drying it shall be subject to all-sliming depending on the operational needs. This is done with a jet-type mill. It is fed with dry products (with humidity which shall not exceed 0.5–1 %). That is why the process includes the drying procedure.

The general flow chart of the pilot production developed by All-Russian scientific-research institute sieve is given on Fig. 4.1. A more detailed production process flow scheme with all separators is given on Fig. 4.2.

This scheme includes the sequence of shungite rocks passing through all process vessels available at this enterprise. The provided flow chart is built in such a way that it provides the obtaining a great number of products which belong to different coarseness categories. The products are passed between the vessels with a helical conveyor.

So, the technology of preparing and processing of the shungite substance includes two stages of crushing, specification, drying and all-sliming of the crushed product. After processing the chemical so as physical and mechanical composition of the original shungite rock practically remains the same.

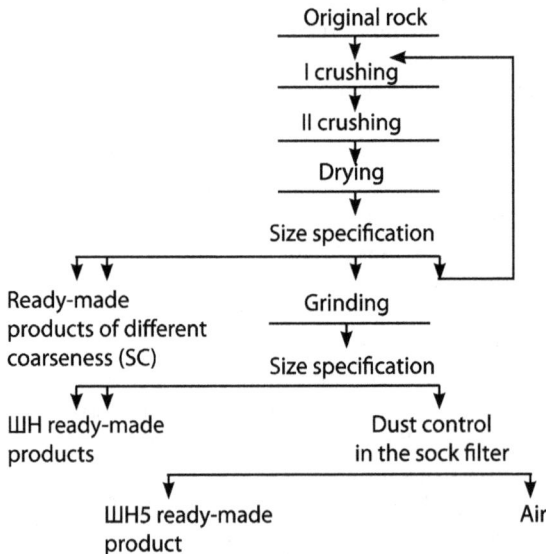

```
                           Original rock
                                ↓
                           I crushing
                                ↓
                           II crushing
                                ↓
                           Drying
                                ↓
                         Size specification
        ↓  ↓                    ↓        ↓
   Ready-made              Grinding
   products of different       ↓
   coarseness (SC)        Size specification
     ↓  ↓                              ↓
   ШH ready-made               Dust control
   products                    in the sock filter
        ↓                              ↓
   ШH5 ready-made                    Air
   product
```

Fig. 4.1. Flow chart of shungite products manufacturing

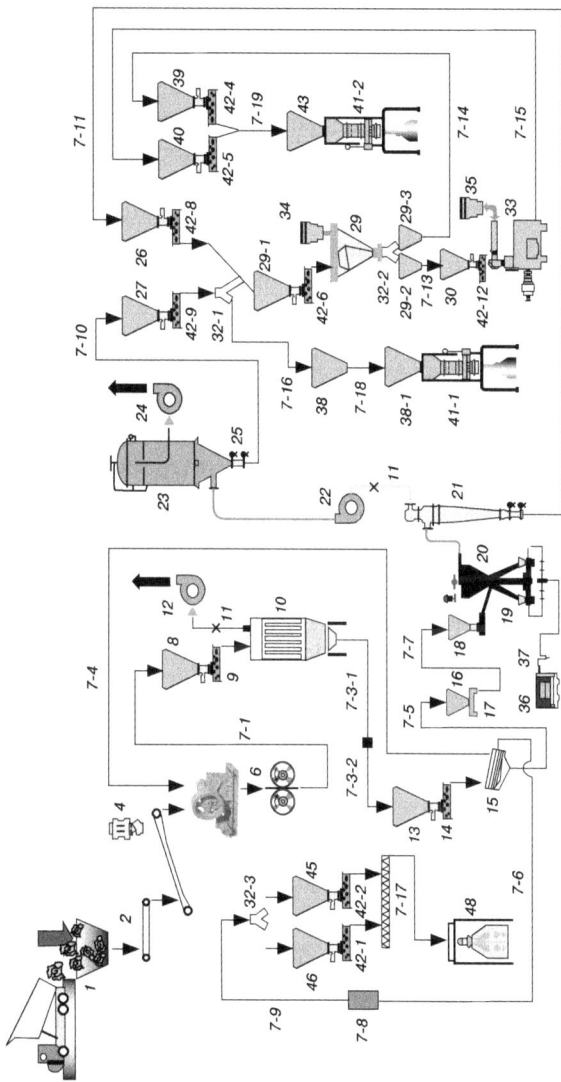

Fig. 4.2. Process flow scheme of the pilot line for producing disperse shungite products:

1 — hopper for receiving of raw materials; *2* — 360/215×940-25 ПЭВ electric-vibrating feeder; *3* — inclined belt conveyor; *4* — ПМЦ 5056 pulley guard magnet; *5* — СМД-116А jaw crusher; *6* — СМД-507 crusher rolls; *7* — SP-90 helicity tube conveyor (Ø90 mm); *7-3-1* — TV-114 horizontal spiral conveyor; *7-3-2* — TV-114 inclined spiral conveyor; *8* — kiln storage hopper; *9* — SF-0,2 screw feeder; *10* — ПЭВ 270 kiln; *11* — nozzle with target; *12* — exhaust fan of ВР-300-45-2Ж drier; *13* — storage hopper for dry crushed products; *14* — ПВ-0,2 screw feeder; *15* — СВ2-0,75/2 sieve separator; *16* — storage hopper for underscreen fractures; *17* — ПРШ-0,2 reversible helix screw; *18* — hopper for raw materials at the plant with the jet-type mill; *19* — counterflow jet-type mill; *20* — classifier of the jet-type mill; *21* — cyclone group of the plant with the jet-type mill; *22* — fan of ВР-132-30-6,3-0,2 mill; *23* — sock filter with ФРИГ-72-type impulse blowout; *24* — fan of ВР-132-8-01 filter; *25* — Ш5-20-РНУ-01 floodgate; *26* — storage hopper for cyclone products (ШН20); *27* — storage hopper for filtered product (ШН5); *28* — reserve; *29* — ПШ-1 agitator; *29-1* — hopper; *29-2, 29-3* — discharge hoppers; *30* — storage hoppers for mixtures ШН5 and ШН20; *31* — reserve; *32* — slide-type Т-junction with DVA-type electric drive; *33* — РВМ-45 vibration mill-modifier; *34* — БПВ-0,35 spiral storage hopper (1 version); *35* — БПВ-0,35 spiral storage hopper (2 versions); *36* — ШТОРМ 2000 spiral compressor (2 pieces); *37* — oil separator; *38* — hopper for ШН5; *39* — hopper for clean or modified ШН20; *40* — hopper for ШН5 and ШН20 mixture without additives or for modified ШН5 and ШН20 mixture; *41* — МФ-type bagger machine for packing of ШН into bags; *42* — screw feeders; *43* — intermediate hopper; *44, 45* — hoppers for chip-adsorbing agent; *46* — machine for packing of the chip into big bags

It is planned that the original rock will comply with all target consumers' requirements towards carbon (25–40 %) and silica content (50–70 %). In fact, this chemical composition of shungite products is the main reference point of developing of the deposit.

We would recommend a foaming apparatus plant widely used in dry metallurgy for treating resurgent gases additionally to the considered project for removing of finely-dispersed dust particles of shungite rock and sulphurous gases from the air formed during grinding and drying. The foaming apparatus plant required the arranging of a closed-looped water rotation with neutralizing of the dissolved gases (refer to Fig. 4.8).

During the process of closed-looped water rotation dissolved sulphur dioxide is precipitated in form of plaster, due to lime impact, $(Ca_2SO_4·2H_2O)$ and filtered and the water cleaned from plaster returns to the system of closed-looped water rotation.

After each crushing, drying, size specification and further all-sliming the obtained fractures are used as final commodity products. Their volume will be determined by pre-defined consumers.

On the first stage of implementing it is provided that granulated shungite products with 1–30 mm crushing coarseness will be manufactured so as finely-dispersed products with coarseness from 5 to 70 μm.

It has been determined experimentally that crushed granulated and fine shungite products can be used without prior mineral processing. The content of carbonaceous matter in such products shall be 25–45 %. Such chemical products can be applied in cast iron, steel and non-ferrous dry metallurgy, construction and chemical industry so as in producing of industrial-rubber products.

For low-grade shungite rocks with 0.5–25 % of carbonaceous matter the intended use is completely different and depends on the chemical composition and rock structural types. These types shall be widely applied in several fields due to their complex development.

4.2. Usage Sphere of Shungite Rocks with Various Coarseness

Before developing and substantiation of the flow chart of manufacturing of shungite products their application in several fields shall be thoroughly analyzed. Practice has shown that for each field shungite products with certain coarseness range and chemical composition are required. As the mineral processing of shungite rocks has not been considered and different consumers require products with different coarseness, the task is to systematize the manufactured shungite products of different coarseness and determine their intended use.

When the flow chart of shungite products manufacturing was developed their physical-chemical composition was not considered as the processes of crushing, grinding and drying of products with different chemical composition are practically the same.

As the flow chart of manufacturing shungite products is determined by the corresponding coarseness category for each consumer all products have to be systematized in compliance with their intended use and required particle size (Table 4.1).

As we can see from Table 4.1., each coarseness category has its own intended use. This gives us the possibility to certify the issued products and spell out its intended use.

Table. 4.1. **Field Application of Shungite Products with Various Coarseness**

Products	Size, mm	Intended use
Slabs	10000–100	Construction, producing of slabs for floors, walls etc.
Coarse rock	100–10	Blast-furnace production of foundry-iron and blast-furnace ferro-alloys
	40–10	Non-ferrous (Co, Ni, Cu), ferro-alloy, white phosphorus electrometallurgy
Stone chippings	10–5	Material for treating of water in wells; filler for radio-shield concretes and conductive asphalts.
	5–1	Velocity filters in treatment facilities
Sand middlings	5–3	Material for water treatment filters
	3–1	
	5–1	Filler for bricks, matrices for masonry and plastery
	3–1	Filler for runner clay
Powder	1–0.7	Filler for industrial-rubber production (ШН70, ШН75 grades); filler for anti-adherent and conductive paints; filler for composition materials for making them antistatic
Finely-dispersed powder	0.7–0.5	ШН50-grade filler; adsorbing agent for wastewater treatment
	0.5–0.25	Filler of industrial-rubber goods (ШН25, ШН5 grades)
	0.5–0.25	Fertilizers
	0.25–0.15	ШН20 filler for rubbers; pigment for construction paints; shungite cement
	0.15–0.05	Aesthetic medicine (pastes, ointments, infusions); ШН5 filler

According to this table, the flow chart of manufacturing of shungite products should provide production of construction slabs, coarse rock and stone chippings, sand and finely-dispersed powder. All this range of products shall be manufactured at enterprises processing the shungite rocks.

As the original shungite rock transferred from the pit has 210 mm coarseness it should be crushed at least with two stages. A standard jaw crusher is used on the first stage and the crusher rolls which are usually used for crushing coal are applied on the second stage. After two stages of crushing a product with 4 mm particles shall be obtained. It shall comprise approximately 90 % of the original shungite rock. For simplifying the crushing technology and reducing dust formation the plant of both grinders on one vertical line with one base shall be used. Such a scheme simplifies the arrangement of structures of dust protection tubes and reduces dust formation. This is very important for operating the crushing plant with a high dust level. After crushing the shungite rock the obtained products shall be classified at the vibratory sieve and particles with the certain range of coarseness depending on the requirements of consumers shall be formed.

The crushed product shall be dried previously and transferred for all-sliming to the jet-type mill which also allow obtaining of finely-dispersed products with a certain class of coarseness. This depends on the operating mode of jet-type mill and specification of the consumer.

After grinding and specification finely-dispersed fractures shall be used as final commodity products with the volume set by the consumers.

Let us consider the application of all proposed products.

Broadstone bind. Broadstone bind or sawn plates shall be used for houses or other industrial buildings constructing, especially when producing electromagnetic shielding. Shungite rocks can be sawn in separate premises. The volume and price of this product have not been determined yet. The shungite outlet and price will be defined and specified within the process of mastering the production and collective business activity. It will depend on the value of the produced plates, quality of their polish and technical finishing-out.

Pit gravel. In foundry-iron and blast-furnace ferro-alloy production of coke and quartz rock are replaced with shungite rocks. The coarseness of such shungite shall be 10–100 mm. The shungite rock with such coarseness is used in producing of foundry-iron and steel-making iron, non-ferrous metals and ferro-alloys. Due to its amorphous structure carbonacous matter is rather resistant to graphitization. Within metallurgical processing it preserves its high reactive and oxidation-reduction properties almost under all temperatures applied to melting and recrystallization in metallurgy. As a substitute of coke and quartz rock due to its stoichiometrical composition shungite rock provides reduction processes in the Si—C—O system and synthesis of metallurgical silicon and silicon carbide as the content of main components in shungite rock — carbonacous matter and silica — is correspondingly 30 % and 60 %, i.e. 1:2.

The mechanism of the impact of shungite rock in this process has not been studied yet, but the first practical knowledge about their application has shown a high viability of these products both technologically and economically. The impact of the content of carbonacous matter and silica on the temperature recovery in Si—C—O system is also unclear. These questions need to be studied specifically and will be specified within commercial exploitation of shungite products at various dry metallurgy plants.

Within metallurgical processing shungite rocks positively influence the kinetics of reduction reactions in the Si—C—O system catalyzing the process. This allows the obtaining of better cast iron. Smooth distribution of silica in the carbonaceous matrix provides an advanced (up to 20 m^2pg) contact between the carbonaceous matter and silica during metallurgical processing. This makes solid-phase reactions more important and accelerates reduction processes. Application of shungite products instead of metallurgical coke and siliceous raw materials is especially viable for obtaining such products as silicone carbide, silicious iron and ferro-alloys. Testing trial batches at several plants has proven that under temperature 1250 °C the active reduction processes begin in shungite products and silicone carbide is formed under 1500—1700 °C. Under 750—1800 °C the whole melting bulk is sharply reduced by 50—60 % in the melting furnace, specifically the silicone carbide (SiC) portion is increased more than by 80—85 %.

The preliminary results of industrial testing show the viability of using of shungite products in metallurgy, especially in blast-furnace melting. Usually for increasing the silica content in cast-iron a set volume of shungite products shall be uploaded into the blast-furnace instead of widely used ferrosilicon.

The shungite product is viable for using in melting in runner and taphole clay as an additive instead of silicone carbide and lump coke. Shungite products positively influence the process of melting blast-furnace ferro-alloys, ferrosilicon, silicon manganese, calcium silicon and ferrochrome silicon in electric furnaces. They also can be applied as a filler and reducing reagent for producing of silicon carbide with further processing into fire-resistant and chemically stable engineering materials.

Firstly, industrial samples for shungite rocks testing during smelting of the foundry-iron were prepared at Carbon-Shungite RPC LLC enterprise under supervision of Y. K. Kalinin and the metallurgical fusion was performed at Tulachermet JSC, Novolipetsk Steel OJSC, Kosay Gora Iron Works OJSC and Cherepovets Steel Mill OJSC. It was determined that 85 % of silica in shungite rocks could be easily transferred into cast-iron. Usually when the silicon content in the cast iron increases the ratio of replacement of coke for shungite product also increases. In average, it will reach 50 %. Usual fusion of steel-making iron provides average consumption of shungite product which comprises 20–25 kgpt of cast iron. In case of smelting of foundry-iron the total consumption of shungite product may increase to 100 kgpt of cast-iron.

It was determined that for blast-furnace melting of ferro-alloys the ratio of replacement of coke with shungite product might increase to 50 %. For silico-manganese melting in electric furnaces the total consumption of shungite product is much lower and is approximately 20 %.

In dry metallurgical processes of producing nickel, cobalt, copper using of shungite products was also appraised due to its high electrical resistance. This allows to perform the fusion with increased content of carbon. High density of the liquid-alloy (2.3–2.4 gpcm^3) provides weaker acidifying of furnace gases. The shungite product has a homogenous structure so it contains much less foreign matter if compared with coke. This improves the quality of the output commodity nickel product because usual additives of coke and quartz rock contain many mixtures (zinc, lead, tellurium, etc.).

The shungite product which is added to the dry metallurgical process contains a little amount of mixtures (strontium, vanadium, gallium, molybdenum) which practically do not affect the quality of the output product.

The shungite product have also been tested on slags of Pechenganikel Mining and Smelting Works JSC, where it has been determined that using it the level of recovered gallium in the concentrated matter can be increased by 23.7 %, nickel — by 5.2 %, copper — by 8.0 %.

Shungite products are good construction materials. They are successfully used for preparing of special concretes as pit gravel or substitute of arenaceous quartz. The coarseness of such product shall be 0.5–2.0 mm. Due to the bloating ability of the shungite product it can be used as a filler for producing of light weight concrete. By burning shungite products an artificial porous material called haydite can be produced.

The shungite product has a high conductivity, mechanical density and durability. It has a perfect adhesive ability for all binding components.

Shungite is used for producing different types of conductive paints, bituminous concrete, a wide range of conductive and radio-shield construction materials.

Conductive shungite paints are eco-friendly and do not educe any harmful substances even when heated. So, such paints shall be used for creating infrared heaters with low rated capacity (from 1 to 10 Wpdm^2). Usually such heaters are safer and cause less fires or burns. It was determined that the infrared heating of buildings is the most cost-efficient one. The total consumption of electrical energy and the losses on maintenance of process equipment are reduced. The main economic effect from infrared heaters is that the consumer received heat without intermediary heating media due to the reflective power of walls, ceilings, floors. This allows reducing the design temperature by 2–3 °C. Conductive shungite asphalts can also be used for heating of warehouses and outdoor freeze-protected sites.

Applying of shungite products as a filler for rubbers instead of white and technical carbon appeared to be especially viable. In this case the coarseness of shungite products shall be 5–75 μm. Such coarseness can be reached when grinding of shungite chip, for example in a counterflow jet-type mill. Experiments proved that the counterflow jet-type mill could produce a very fine shungite product with a narrow range of coarseness (e.g., ШН5 grade — (5–10 μm); ШН20 grade — (15–20 μm); ШН75 grade — (50–75 μm).

Currently, manufacturing of shungite products has been developed in Karelia at Carbon-Shungite RPC LLC enterprise under supervision of Y. K. Kalinin. This enterprise mines and processes shungite rocks making various products (refer to Table 4.1).

The intended use of shungite products according to technical conditions is given in Table 4.2 For extending the intended use of shungite products additional industrial facilities shall be constructed on the base of Karelian shungites. The works in Lianozovo district, Moscow, can serve as an example of such facility (LLC "MKK-Engineering", manager — V. A. Maslakov).

After commissioning of this facility it is assumed to extend the stock list of the manufactured shungite products and execute additional specifications to their production. Similar engineering solutions are required for manufacturing of products used for making black for paint industry, electronic industry, etc.

High-grade finely-dispersed shungite products can be also used in agriculture as fertilizers. The first experiments for applying such fertilizers were successful [Nartov, 1798; Pekki, 1975].

Table. 4.2. **Stock-list of ready-made products**

Product name and intended use	Technical conditions, grade	Main technical data				Non-toxic, low-combustible material
Crushed middlings-drying agent for local water treatment systems, reduction-alloying additive for ferrous metallurgy	ШК СТП21-04604485-97	Content of particles larger than 4 mm does not exceed 3 %, content of particles smaller than 0.5 mm — up to 3 %. Humidity — up to 1 %. Carbon content — at least 25 %				Non-combustible, non-toxic
Finely-dispersed organomineral active filler for elastomerss, thermoplastics and thermosets, etc. ШН	TU 2169-002-5684-0806–2003		ШН5	ШН20	ШН75	Flammability category — low-combustible material, ignition point — 480°C; powder does not cause inflammation; LFL of the aerial suspension of ШН5 and ШН20 mixture — 175 gpm³ in case of presence of the igniter with temperature exceeding 1000°C, i.e. the explosiveness of ШН5 and ШН20 is mild.
		for carbon (in %)	25–40	25–35	25–35	
		humidity (in %)	up to 0.5	0.5	0.5	
		pH of the aqueous extract	5–5.7	5–7.5	5.75	
		specific surface (in m²pg)	25–40	10–25	5–15	
		loose density (in kg²pdm)	0.2–0.3	0.4–0.6	0.5–0.7	
		residue (n %) after screening through a meshy sieve	0.14	–	0.5	
			–	0.005	0.5	
			0.045	0.001	4.5	

Currently finely-dispersed shungite product comes into use for medical purposes for making creams, masks, shampoos. It is viable to apply it in medicine as a unique germicide [Orlov, 2004]. Aqueous infusions of shungite products have medicinal properties and can be used for treating of many diseases: allergic, skin, respiratory, muscle diseases. Also, they have antipruritic and antiphlogistic properties [Rysyev, 2004]. It is assumed to use shungite drugs in cosmetics and pharmacology in future.

Particles of finely-dispersed shungite powder are bipolar so they can mix well with all known chemical substances and with fluoroplastics, rubbers, resins, cements, solvents. This allows obtaining of different combinations and reaching a maximum technological and economical effect.

Mixing shungite powders with butadiene-methylstyrene, isoprene and divinyl rubbers with 3:100 ratio provides good handling abilities of the wire-frame rubber mixture which can be used for making of high-quality road rubber tyres. It was determined that shungite powders create less dust during mixing and can be well mixed with rubber. This allows improving of the sanitary and hygienic conditions at the plant.

Shungite products were also tested during making of rubbers for footwear. Usually the shungite powder is added into the styrenebutadiene rubber mixture for facing instead of П-803 technical carbon. This mixture allows improving physical-chemical properties of rubber products by 30 %. The Moony viscosity of this mixture is reduced twice and the vulcanization period is increased. The durability is increased by 10 % and rubber elasticity is improved by 20 %.

It was shown that substituting of T-900 technical carbon by shungite powder practically had not changed the main handling abilities and structural behaviors of the obtained rubbers when producing of conveyor belts.

Shungite products were applied as highly active drying agents when cleaning waste water from oil products and heavy metals [Kalinin, 2002; Kalinin, Pekki, 1997].

We deem appropriate to begin implementing of shungite products at velocity filters of main treatment facilities at Kuryanovskaya and Lyuberetskaya aeration station in Moscow replacing arenaceous quartz with shungite chip. For velocity filters of Moscow shungite chip with coarseness of 1–5 mm shall be used. As it has been already mentioned, the shungite product improves the sanitary and hygienic properties of water by reacting with it. Besides that, possessing catalytical and cation-exchange characteristics it acts as an adsorbing agent reducing the content of phenols, guaiacols, eugenic and oleinic acid in the treated water.

A great diversity of particle coarseness in applied shungite products makes the technology of their production more complex and requires a corresponding technological scheme.

Due to the fact that the average humidity of original shungite rocks from the pit is 8 % they should be pre-dried before all-sliming.

So, on the basis of the analysis of the obtained shungite products and their physical-chemical properties the simplest and cost-efficient dry technology of processing has been adopted. The water at the factory is used only for cooling compressors and for hygiene needs.

4.3. New Areas of Complex Usage of Shungitous Rocks

Due to complex processing of shungite rocks a range of new products has been found out which can be obtained from different types. Let us consider the most viable products which should be manufactured from low-grade shungitous rocks with processing in the steam chamber, chemical leaching and burning.

The flow chart for processing of high-grade and low-grade shungite rocks includes procedures which are common for them only at the preparatory stage. Those are crushing, size specification, drying and grinding. On the first stage this flow chart can be used for processing only of high-grade rocks with 25–45 % of carbonacous matter with obtaining of products of different coarseness (refer to Fig. 4.1).

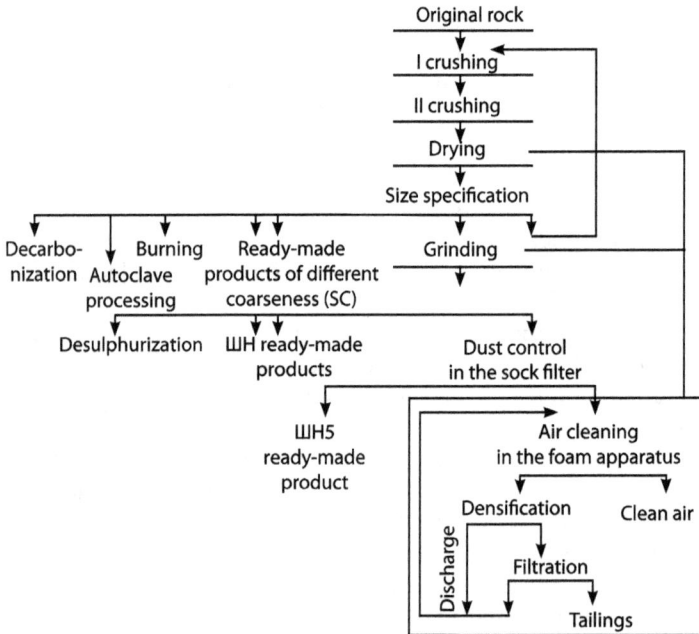

Fig. 4.3. Advanced flow chart of manufacturing of shungite products

For shungitous rocks with 0.5—25 % of carbonacous matter new operations shall be added into the production scheme: leaching, burning, processing in the steam chamber.

The new flow chart provides cleaning the exhaust air from solid particles and sulphur-containing gases so as a system of closed-looped water rotation and waste water treatment. Obtaining of new products with leaching, burning and processing in the steam chamber shall be performed separately.

A complex approach to processing shungite rocks has shown that each type of low- and high-grade rocks has it intended use after corresponding technological improvement. E.g., potassium silicate rocks can be used for producing of liquid glass after processing in the steam chamber and aluminosilicate rock serves as good raw materials for producing haydite while as carbonate types of low-grade rocks are a source of lime after burning. The mixture of aluminosilicate and carbonate rocks is the best raw material for ordinary cement.

All these products are made by burning, processing in the steam chamber and chemical leaching. These processes are described briefly below.

4.4. Burning Technology for Obtaining of Some Products from Shungite Rocks

Burning technologies are applied to shungite rocks for producing of lime, haydite, cement, calcium carbide. For obtaining of such products shungite raw materials with different chemical composition should be used. Each product requires a certain temperature and set chemical composition of the original raw materials.

For lime carbonate rocks producing with low content of shungite which contain up to 25 % of carbonacous matter and at least 30 % of carbonates should be used. The temperature of burning shall be 800—900 °C.

The scheme of lime obtaining is given on Fig. 4.4 a.

Haydite can be obtained from aluminosilicate rocks with low content of shungite and 5—10 % of micaceous minerals, used for thermal insulation in construction (Fig. 4.4 b).

Aluminosilicate-carbonate rock with low content of shungite and clay and hassock admixture is a raw material for cement which shall be produced in a rotating drum furnace 150—170 m long under firing temperature 1600—1700 °C (Fig. 4.4 b).

As one can see all technologies shown above have a different burning temperature. Besides that, shungite rock with different chemical composition is used for each product.

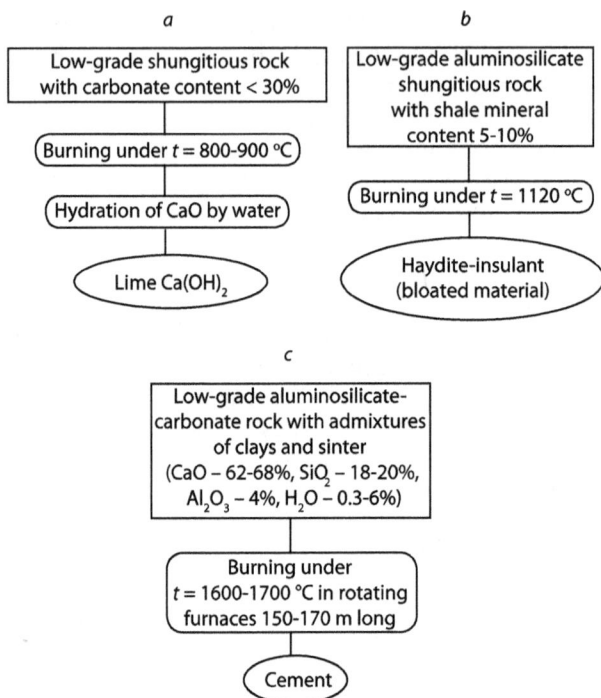

Fig. 4.4. The technology of shungitous rocks burning using heat received by burnoff of carbon: *a* — production of lime; *b* — haydite; *c* — cement

The technologies are economically viable due to using heat from the burnoff of the carbonacous matter in shungite rocks.

Carbonate rocks with low content of shungite are of interest not only for obtaining lime, but also for high-grade rocks for reducing their acidic properties. For this carbonate rocks with low content of shungite shall be added to the main bulk of shungite rocks with acidic medium for neutralizing the mixture. Acidic constituents are neutralized according to the reaction:

$$CaCO_3 + H_2SO_4 = CaSO_4 + H_2O + CO_2\uparrow.$$

Carbonate shungitous rocks with low content of carbonacous matter which contain more than 40 % of carbonates can be used for producing lime and calcium carbide. In this case the firing temperature shall be at least 800—900 °C. During the burning process the carbonate constituent of shungitous rock is decomposed according to the reaction:

$$CaCO_3 = CaO + CO_2\uparrow.$$

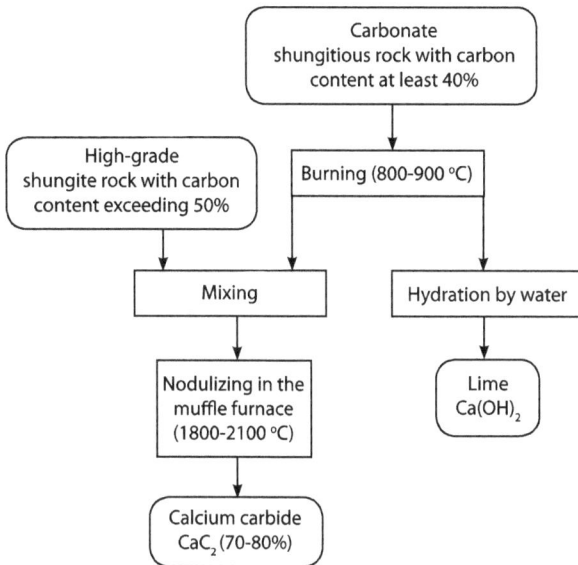

Fig. 4.5. Process design of calcium carbide and lime

During burning the carbonacous matter in the rock (1–15 %) is fully burnt off creating additional heat which is used for burning. Calcium oxide which is formed during this is used as lime. As calcium oxide further reacts with water lime is formed. It can be used as a construction material.

The carbonate rock with low content of carbonacous matter which contains 43 % of carbonates was thermally processed for 2 hours in the muffle electric furnace under temperature 1000 °C in laboratory conditions. During the burning process a small amount of carbonacous matter (8 %) was burnt-off and the formed calcium oxide was a typical grey porous powder. When the calcinated product is treated by water it dissolved and a deep milky lime solution with pH 12.5 is formed.

When the calcinated carbonate rock with low content of shungite is agglomerated in the high furnace mixed with high-grade shungite rock calcium carbide is formed. The process design of calcium carbide and lime is given in Fig. 4.5.

Haydite production is one of the most important and wide-spread areas of applying aluminosilicate rocks with low content of shungite. Aluminosilicate rocks with low content of shungite and lyddite and mica admixtures are good raw materials for producing expanded clay. To obtain expanded clay the average content of micaceous minerals in rock shall be at least 8–10 %. Under temperature 1120 °C pelitic and micaceous minerals in such aluminosilicate rocks are recrystallized while the heated bulk is bloating. The level of bloating depends on the content of micaceous minerals and the volume of the shungite product can increase more than 5 times.

Usually for producing good expanded clay raw materials with $K_{blt} = 2,5$ (GOST 9757-90(2002) — "Artificial porous gravel, crushed stone and sand.") bloating coefficient is required.

When finely-dispersed aluminosilicate rocks with low content of shungite are burned pyroplastic phases are created which bulge rapidly when reacting with micaceous admixtures and form a new material called haydite. During burning of shungitous rock the carbonacous matter is fully burnt-off and the heat created by this process increases the temperature of the calcinable material. In such case most minerals including clinkstone and hydromicas are oxidized creating new phases. Carbonate admixtures in the raw materials negatively influence the bloating coefficient. Their impact can be explained by high chemical activity of calcium, magnesium and iron oxides which rapidly react with water after burning and new constituents unacceptable for haydite are formed. So low-grade rock without carbonates is used for making haydite.

In the process of burning and bloating argilliferous and lyddite-containing rocks at Zazhogino deposit light and durable pellets, which remind common expanded clay by their physical-chemical and technological properties, formed. The technology of obtaining such expanded clay from aluminosilicate rocks with low content of shungite is similar to the previously used technology [Borisova, Klimov, 1974].

For checking the possibility to obtain the expanding clay test burning of aluminosilicate rocks with low content of shungite in the muffle furnace was held. The sample has the following content: SiO_2 — 38.5 %, Al_2O_3 — 40.1 %, Fe_2O_3 — 8.3 %, CaO — 1.5 %, C — 5.3 %. The temperature of heating the material in the furnace was 1150 °C. The time of temperature-conditioning was 2 hours.

The calcinated material bloated, the bloating coefficient was 2.9. This proved the possibility to obtain haydite from aluminosilicate rocks with low content of shungite. During the burning process the carbonacous matter fully burnt-off and the produced heat was used for increasing the temperature of the roast mixture. If this process will be developed this heat will allow reducing the total expenses on energy carriers and thus the cost price for the product.

The volume of aluminosilicate rocks with low content of shungite at Zazhogino deposit which may bloat during burning is enough for their industrial use. So specific works for determining the chemical content of the rocks best for haydite production are required after constructing the industrial facility and mastering the production.

The most complex technology of burning shungite rocks is required for cement production. It is one of the main construction materials. Everybody knows that its production is one of the most difficult tasks in construction. Currently there are lot of types of cement made at different plants of the Russian Federation. Usually the grade and type of cement depends on the original raw materials, level of rock crushing and method of production and firing temperature.

Rocks with low content of shungite of Zazhogino deposit can be used as one of the sources of raw materials. Their carbonacous matter will burn under temperature which exceeds 1000 °C and the heat will reduce the loss of energy carrier and thus the cement cost price.

Before considering cement properties obtained from aluminosilicate-carbonate rock let us consider the main peculiarities of wide-spread cements made of standard raw materials used at most of the existing Russian cement works.

Usually for obtaining high-quality portland cement the original raw materials for standard raw cement clinker shall contain the following components: CaO — 62–68 %; SiO_2 — 18–26 %; Al_2O_3 — 4–9 %; Fe_2O_3 — 0.3–6 %. These data are known from good practice and average reports of cements works,

Besides that, all raw materials with a high content of calcium oxide and aluminum silicates may be used for producing cement. The most wide-spread carbonate aqueous rocks usually satisfy these conditions. Besides carbonates and pelitic minerals, the rock shall also contain finely-dispersed silica and a small number of iron oxides.

Usually the ratio between clay and carbonates in rock shall be 1:3. In the cement mixture limestone can be replaced by marl, chalkstone or carbonates in the shungitous rock. Aluminosilicate-carbonate rock with high or low content of shungite from Zazhogino deposit fully meets the requirements for the chemical composition for the raw mixture for producing standard cement.

All necessary additives shall be put into the raw mixture if it does not meet the standard specifications.

The obtained raw mixture shall be burned and the calcinated material forms the standard cement.

To conclude it should be mentioned that all types of rocks with low content of shungite (silicate, aluminosilicate, silicate-sodium, aluminosilicate-carbonate, carbonate) have their peculiar technological properties and intended use. These rocks can be used for obtaining many chemical products for use in industrial practice: sodium silicate, calcium chloride, calcium metal, calcium carbide, lime, cement and other products.

4.5. Shungite Rock Processing in Steam Chamber for Liquid Glass Producing

One of the most interesting problems in the sphere of using shungites is production of liquid glass (technical name — "water-glass"). It shall be made from rocks with high content of silica and potassium-containing minerals. There are quite a lot of such types among low-grade shungite rocks at Zazhogino deposit.

These types of rock have not come into industrial use yet, although they are very viable. For starting industrial manufacturing of liquid glass investments from builders, ore enrichment specialists or chemists are required as currently there is a lack of producing liquid glass and its substitutes are absent.

Within the last 50 years many types of potassium silicates have been introduced into industrial manufacturing. They are used for producing liquid glass with various content of sodium oxides and silica. Those are sodium orthosilicate Na_4SiO_4, or $2Na_2O \cdot SiO_2$; sodium metasilicate Na_2SiO_3, or $Na_2O \cdot SiO_2$; sodium disilicate $Na_2Si_2O_5$, or $Na_2O \cdot 2SiO_2$; sodium trisilicate $Na_2Si_3O_7$, or $Na_2O \cdot 3SiO_2$.

At the moment the mechanism of producing liquid glass has been elaborated already [Matveyev, 1957]. Its concept can be represented as the chart of fusibility of $Na_2O \cdot SiO_2$ silicates (Fig. 4.6).

The given chart contains all above-given sodium silicates (Na_4SiO_4, Na_2SiO_3, $Na_2Si_2O_5$, $Na_2Si_3O_7$). Rocks with low content of shungite often contain such compounds and may be found at Mironovsky-1 section. Sodium silicates differ in the ratio of sodium oxide and silica. Their fusing point depends on the silica content. Sodium trisilicate has a higher fusing point because it melts under 1600 °C and higher. This substance has more than 75 % of silicon.

It is more viable to smelt sodium disilicate $Na_2Si_2O_5$, its fusing point is 800–850 °C and the ratio between sodium oxide and silicon is approximately 1:2.

If there is a very high content of sodium oxide and low content of silicon (appox. 40–60 %) the fusing point of the mixture increases to 1100–1150 °C and if the content of silicone oxide is reduced (less than 40 %), it will be impossible to produce the required liquid glass.

Using the developed scheme, it is easy to determine the dependency between the behaviour of different sodium silicates and the fusing point. Thus, under 1115 °C sodium orthosilicate Na_4SiO_4 is smelted incongruently educing Na_2O and silicate fusion.

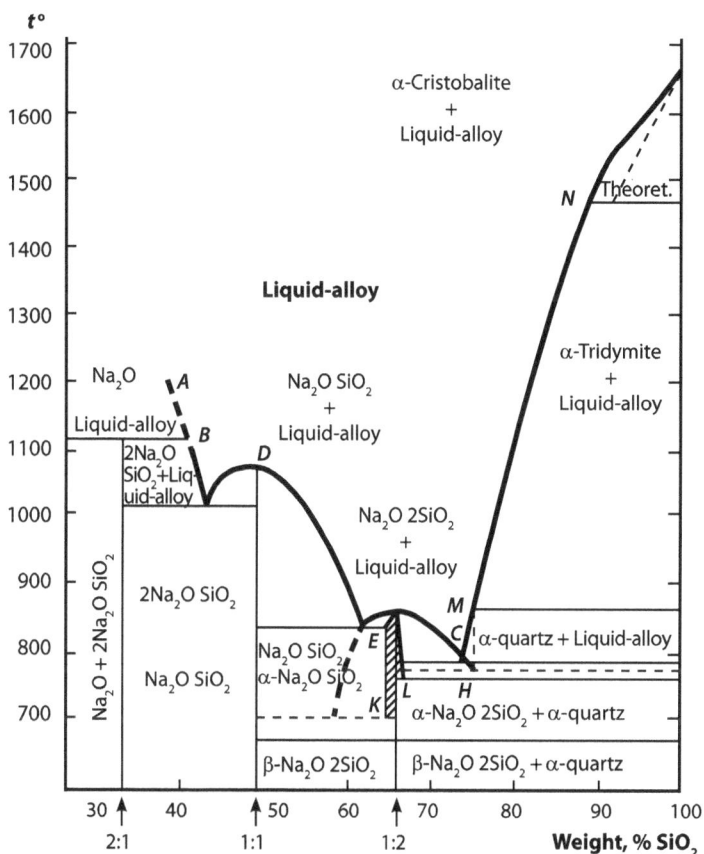

Fig. 4.6. The chart of fusibility of Na_2O–SiO_2 system in shungite rocks

Na_2SiO_3 is also incongruent under a certain set temperature. Sodium disilicate is at the highest point on the fusibility curve. This phenomenon is related to the dissociation of the formed compounds.

Considering this chart, presence of three eutectic points is obvious. The first one is related to forming of sodium orthosilicate under 1020 °C; the second one — with forming the mixture of sodium metasilicate and disilicate under 846 °C; the third one — with smelting sodium disilicate under 800 °C. Silica crystals are educed from this point of fusion. Sodium trisilicate is crystallized from them and dissolved under 1600–1700 °C.

In case of a higher content of SiO_2 the solubility curve goes up when the temperature increases to 1700 °C. This happens due to the third eutectic points $Na_2O \cdot 2SiO_2$–SiO_2 up to full melting of SiO_2.

Along the liquidus line a lot of intermediary solutions with different content of $Na_2O \cdot nSiO_2$ may exist between three silicates and their eutectic. All considered types of sodium silicates (orthosilicate, metasilicate, disilicate, trisilicate) can be obtained from low-grade shungite rocks. For this one has to select the rock with the required original content of sodium-containing minerals and silica. All considered types of sodium silicates are easily dissolved in water forming liquid glass, one of the main chemical reagents which regulates the flotation process.

The problem of producing liquid glass from low-grade shungitous rocks requires additional special researches with further execution of regulations and potential introduction. This paper does not cover such researches.

The flow chart of manufacturing liquid glass — waterglass from silicate-sodium rocks with low content of shungite is rather simple.

The original silicate-sodium rock with low content of shungite and 1−15 % of carbonacous matter is loaded into the steam chamber and processed under 800−1700 °C. The processing temperature is set depending on the ratio between silicate and sodium minerals in the rock which may be different. According to the chart on Fig. 4.6 the best ratio shall be 1−2 while the fusing point in the steam chamber is minimal and is approx. 800−850 °C. If the sodium content in the rock increases the fusing point also increases to 1100 °C. If the silicate content increases the temperature rises to 1600 °C.

To conclude it should be mentioned that the technology of processing shungite rock for producing sodium silicate in the steam chamber is simple and can be easily introduced into industrial manufacturing.

The process of producing sodium silicate can be performed during pilot production in Lianozovo (MCC Engineering Ltd.) or the Republic of Karelia (Carbon-Shungite RPC LLC). Such production requires financing, standard operating procedure and procuring autoclave equipment.

4.6. Production of Calcium Carbide from Shungite Rocks

Shungitous carbonate rock are very viable for producing calcium carbide which is a very raw material used in several fields, especially in metal cutting and welding and metallurgical melting of iron ore and non-ferrous metals.

For producing calcium carbide two types of rocks can be used: high-grade shungite rock with at least 60 % of carbonacous matter and low-grade carbonate one with at least 30 % of carbonates. Before mixing, carbonate rock with low content of shungites shall be subject to preliminary burning for obtaining calcium oxide. After adding water it is turned into caustic lime according to the reaction:

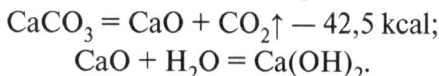

$$CaCO_3 = CaO + CO_2\uparrow - 42,5 \text{ kcal};$$
$$CaO + H_2O = Ca(OH)_2.$$

During the burning process carbonates are decomposed, i.e. dissociated with absorption of a lot of heat (42.5 kcal). When carbonates are decomposed the quality of the produced calcium oxide will depend on the chemical composition of the original shungite rock and the firing temperature. The cleaner is the original carbonate rock, the higher is the quality of the composition of the mixture for producing calcium carbide. Burning of carbonate rocks with low content of shungite shall be performed under 800–900 °C.

Calcium carbide is produced by melting together lime and rock with high content of carbonacous matter in a pit furnace as the calcinated lime is reduced with carbon according to the reaction:

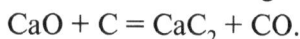

$$CaO + C = CaC_2 + CO.$$

The formed calcium carbide usually contains a high amount of admixtures which do not participate in the process. Calcium carbide content in the obtained product shall be at least 85 %, where CaO, MgO, SiO_2, C act as admixtures.

The obtained product can be easily decomposed in water educing gas (acetylene) according to the reaction:

$$CaC_2 + 2H_2O = Ca(OH)_2 + C_2H_2\uparrow.$$

This reactions takes place in case of water excess. If there is a lack of water the chemical process changes:

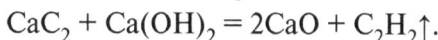

$$CaC_2 + Ca(OH)_2 = 2CaO + C_2H_2\uparrow.$$

Calcium carbide is a very strong reducing agent for most iron oxides: Pb, Sn, Fe, Co, Cr, Mo, Mg, etc.: it is also a strong oxidant in pyrometallurgical producing of steel.

4.7. Chlorohydric Acid Leaching of Low Shungitous Carbonate Rocks

Used for Calcium Chloride Producing

Calcium chloride can also be obtained from carbonate rock with low content of shungite $(CaCl_2 \cdot 2H_2O).(CaCl_2 \cdot 2H_2O)$. The technology of obtaining it provides processing of carbonate rock with low content of shungite with chlorohydric acid with 13–15 % concentration and mixing in the conditioning tank.

Carbonate shungitious rock
with carbon content
0.5-15 %

13-15% solution HCl ⟶

Mixing

Filtration

Calcium chloride
solution
($CaCl_2$)

Residue
piling

Drying (in the flash drier)

Crystallohydrate
of calcium chloride
($CaCl_2 \cdot 6H_2O$)

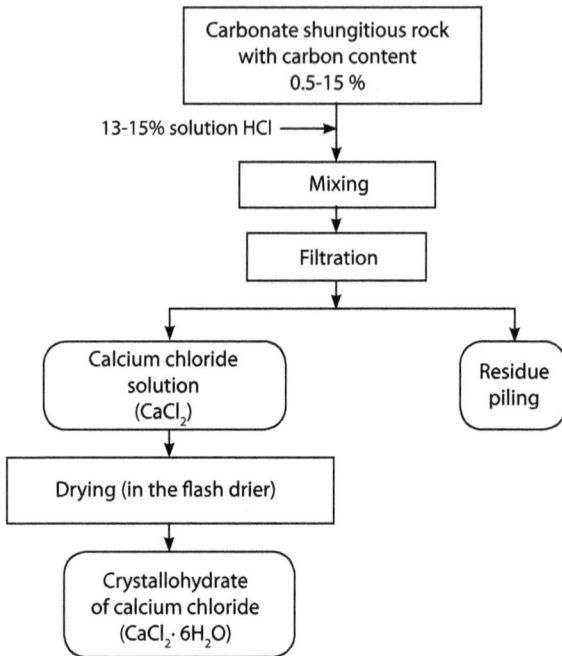

Fig. 4.7. Process design of calcium chloride
from carbonate rock with low content of shungite

We have tested the possibility of obtaining calcium chloride in laboratory conditions on the shungitous carbonate rock with 8 % of carbonacous matter and 43 % of carbonates. This rock was easily dissolving in chlorohydric acid educing CO_2. During dissolution the remaining original weighed quantity reduced 2.7 times. This proves good solubility of carbonate rock with low content of shungite as the content of carbonacous matter in the residue after dissolving carbonates increased up to 25 %. The process design of calcium chloride is given in Fig. 4.7.

The obtained solution of calcium chloride is separated from the solid remainder by filtering. To obtain calcium chloride from the liquid solution it shall be dried in the flash drier. In this case a solid white powder of calcium chloride is crystallized from the solution used as a chemical reagent in different spheres, in particular, in winter road treatment and at mineral processing factories. The non-dissolved residue of shungite carbonate rock can be used as concentrated shungite product.

Calcium metal is obtained electrochemically from calcium chloride solution according to the standard scheme. It is used in metallurgy and calcium silicate is obtained by thermal processing of calcium chloride in electric furnaces.

All above-listed products are rather sought-after and expensive so works on obtaining them from shungitous rock are of great practical and theoretical interest, although researches in this sphere have not been performed yet.

4.8. Products Obtained from High-Grade Shungite Rocks

Activated shungite rock also can be obtained from high-grade shungite rocks which contain more than 40 % of carbonacous matter according to the standard technology. It is widely used in industrial practice, in particular for treating waste water and air.

Besides that, high-quality insulant in form of flexible plates is received fro m high-grade shungite rocks by latex soaking of the finely-dispersed shungite product. The flow chart of manufacturing the activated shungite product and insulant is given in Fig. 4.8.

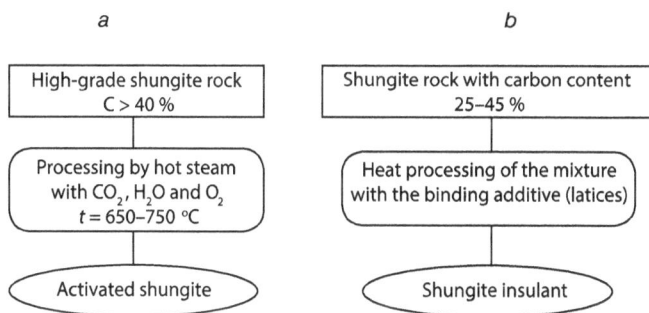

Fig. 4.8. Technologies of manufacturing products
from high-grade shungite rocks:
a — absorbent carbon, 6 — shungite insulant

Activated shungite product from high-grade rocks with 25–98 % of carbonacous matter can replace almost all absorbent carbons produced by native industry.

It is recommended to use the steam-gas method for manufacturing this product based on physical activation. This process provides hot air steam processing the crushed shungite product with coarseness 1–2 mm under 650–750 °C [Fedoseyev, 1963]. CO_2, H_2O and O_2 are present in the steam mixture.

The activation process takes places under 600–700 °C generating heat according to the reaction:

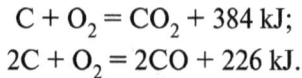

$$C + O_2 = CO_2 + 384 \text{ kJ};$$
$$2C + O_2 = 2CO + 226 \text{ kJ}.$$

When the carbonacous matter reacts with the water steam the following reactions with heat absorption proceed:

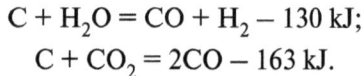

$$C + H_2O = CO + H_2 - 130 \text{ kJ};$$
$$C + CO_2 = 2CO - 163 \text{ kJ}.$$

The speed of these reactions depends on the speed of oxidizing of the carbonacous matter and removing the formed products from the active oxidation zone. As one can see, the process of activation of carbonacous matter has two stages: heat generation and absorption. Both processes allow obtaining high-quality activated products without overheating. Oxides and carbonates of alkali metals and iron and copper present in shungite rock catalyse the processes of activating shungite products with steam generation. We consider that the analyzed method of activating shungite products shall be implemented at the enterprise which is constructed in Lianozovo.

It is assumed that the industrial manufacturing will significantly extend the intended use of several types of high- and low-grade shungite rocks. This will allow solving the problem of complex use of all types of rock and improvement of the enterprise technical and economic performance.

4.9. Application of Shungite Products in Chemical Machine Engineering

G. S. Petrov Research Institute of Plastics OJSC has developed the process of obtaining several composition materials for repair and rehabilitation of the range of process equipment at chemical plants from shungite products which contain 25–35 % of carbonacous matter. "Shungite 251" and "shungite 520" serve as an example of such materials. They are produced by mixing and cold curing of finely-dispersed shungite product using epoxy resin. The technology of manufacturing these products is implemented according to TU 2257-451-00209349-2006. They are used at chemical mechanical engineering, housing and utilities infrastructure, oil processing and cellulose and paper enterprises. The intended use of "shungite 251" and "shungite 520" is producing wearproof coating of pumps, containers, agitators, etc.

4.10. General Description of Enterprises Manufacturing Shungite Products

Production of shungite products in the Republic of Karelia has been organized at two industrial enterprises: Carbon-Shungite RPC LLC and Kondopoga shungite plant LLC.

Both enterprises use shungite rocks from different sections of Zazhogino deposit. Their chemical and mineral composition and structure are different so this provides different intended use.

Manufacturing shungite products has also been mastered at Koksu Mining Company LLP — a large industrial enterprise in Kazakhstan at the premises of the former subdivision of Tekeli polymetallic complex.

There is a range of experimental-industrial enterprises which study, process and manufacture new types of shungite products widely used for household and medical purposes. LLC "MKK- Engineering" (Moscow), Promkontsentrat-Tekhnologii LLC (Naro-Fominsk, Moscow region), Filtry MM LLC (Saint Petersburg), Pritsero-P LLC (Moscow).

Carbon-Shungite RPC LLC This enterprise develops shungite rocks since 1991. Its performance is approx. 200–250 thousand tpy. They perform open-pit mining at two active pits: Zazhogino and Maksovo sections of the deposit. It plans to increase the volume the extraction of the rocks up to 350 thousand tpy.

The enterprise manufactures a wide range of shungite products of different coarseness used in metallurgical, chemical, construction, industrial-rubber and other fields.

Processing operations: crushing, grinding, size specification. The cost of dispatched products depends on their coarseness. Value indicators of dispatched products (according to the data of the enterprise for 2006) are given in Table 4.3 as an example.).

Table. 4.3. Shungite products made by Carbon-Shungite RPC LLC
and their cost

Fracture size, mm	Cost including VAT (in RUB per ton)
1–3	7400
3–5	5900
0–3, 0–5	4500
3–10, 5–10, 10–20	4800
0–10, 10–40	3800
10–100	2500
powder 0–0.02	12000

The production volume at the enterprise increases every year. Its growth for 2002–2007 is shown in Fig. 4.9.

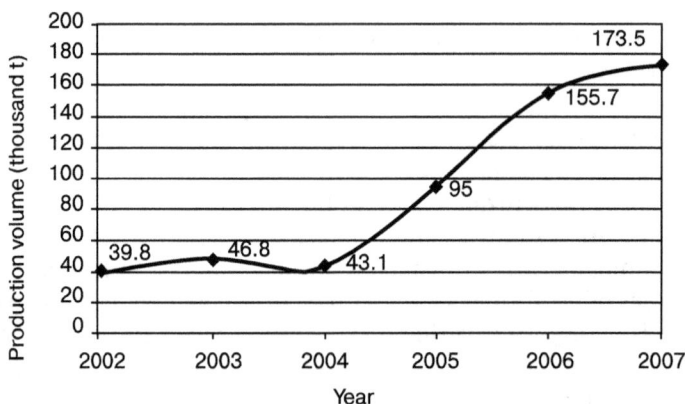

Fig. 4.9. Dynamics of growth of goods production for 2002–2007

Russian metallurgical enterprises using shungite products as coke substitute in foundry-iron production are the main consumers of its products. Enterprises-consumers of shungite products: Kosay Gora Iron Works OJSC (Tula), Svobodny Sokol LIW OJSC (Lipetsk), Novolipetsk Steel OJSC (Lipetsk), Novokuznetsk Iron and Steel Plant OJSC (Novokuznetsk), West-Siberian Metal Plant OJSC (Novokuznetsk), Tulachermet JSC (Tula), Severstal OJSC (Cherepovets, Vologda region) and many other big Russian enterprises.

Additional facilities for manufacturing finely-dispersed shungite powders were commissioned in 2004 near Petrozavodsk. The volume of powder production is 400 thousand tpy, among them — not less than 50 % of finely-dispersed powders. It is planned to supply these powders to industrial-rubber enterprises, including Kvart Kamsko-Volzhskoye aktsionernoye obshchestvo rezinotehniki CJSC (The Republic of Tatarstan), Ecochimmach CJSC (Buy, Kostroma region), Volzhskrezinotekhnika CJSC (Volzhsky, Volgograd region).

Kondopoga shungite plant LLC processes shale rock from Nigozerskoye section of Zazhogino deposit. The works began in 1972 and in 2004 the enterprise facilities were restored.

The enterprise specializes on producing shungite pit gravel used for road construction and concrete production. This pit gravel fully complies with requirements of GOST 8267–93 "Crushed stone and gravel of solid rocks for construction works".

The dynamics of growth of goods production at the enterprise is as follows (Fig. 4.10).

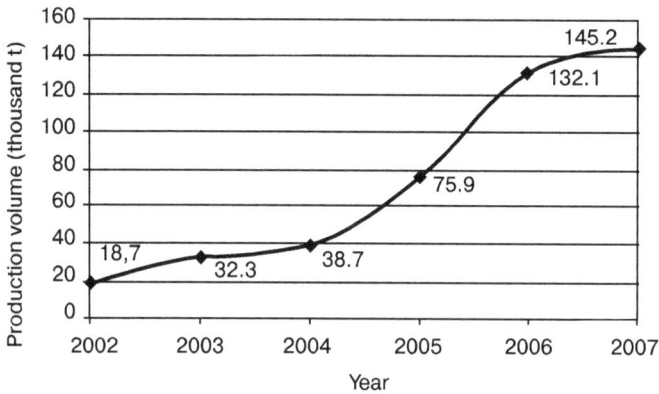

Fig. 4.10. Dynamics of growth of goods production for 2002–2007

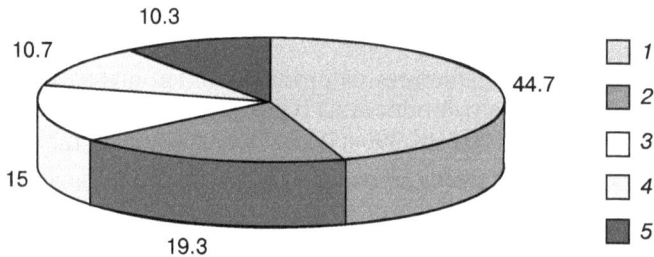

Fig. 4.11. Regional consumers Kondopoga shungite plant LLC (%):
1 — Moscow and Moscow region; *2* — Saint Petersburg and Leningrad Region; *3* — Tver region; *4* — Vologda region; *5* — others

Shungite shale rock from Nigozerskoye section are widely used for producing haydite used for thermal insulation in constructing residential and industrial buildings. It is also used for producing light weight concrete. Recently the haydite production has reduced. The cost of 1 m^3 of shungite shale rock of different size at Nigozero station is as follows: 5–20 mm — 185 RUB.; 20–40 mm — 125 RUB

Kondopoga shungite plant LLC products are used at enterprises in Moscow, Saint Petersburg, Moscow region, Vologda, Tver and Leningrad region (Fig. 4.11).

Enterprises-consumers of shungite shale products: Soyuz Les CJSC (Moscow), Stroy Montazh LLC (Moscow), ZBI-21 JSC (Moscow), TSM-Neman LLC (Saint Petersburg), North-Western Non-Metallic

Company CJSC, Chemical Mechanical Engineering RI FSUE (Moscow region), Stroitel Firm CJSC (Moscow region), Zavod Krupnopanelnogo Domostroeniya branch (Moscow region), Vash Dom LLC, Neman LLC (Leningrad region), Zakneftegazstroy-Promitey (Tver region), ZBI-Komplekt LLC (Tver region), Shungizit OJSC (Murmansk region), etc.

Koksu Mining Company LLP. The structure and features of shungite rock of Koksu deposit are different from shungite rock of Zazhogino deposit. Their main technological features are similar to those of Karelian shungite rock. The deposit is at Rudnichy settlement of Almaty region south from Tekeli on the altitude approximately 2000 m. The deposit has been opened in 1986, its reserves are 620 mln t, among them those which belong to category C_2 — 3.6 mln t. The development is made by open-pit mining with selective incision of shale, carbonate and hybrid rocks.

Crushing, grinding, size specification and flotation are performed at the premises of the former Tekeli polymetallic complex. The capacity for the shungite flotational concentrate of the crushing plant is 1 mln tpy and that of the processing unit is 400 tpy. The process system includes the crusher house, mill with a classifier, flotational shop, densification and filtration shop, desintegration and drying department, ready-made product pre-packing shop, mining warehouse, storage facilities, overhaul shop, etc.

The enterprise manufactures its products under Таурит trademark. Koksu Mining Company produces six types of taurits according to technical conditions TU 1900 PK 39646043 TOO-003—2003 (Table 4.4).

Table. 4.4. **Types of taurits produced by Koksu Mining Company LLC**

Name	Grade	Size distribution
Shale taurit	TC	0–35
Shale taurit (after desintegrator)	ТСД	up to 100 μm
Flotational shale taurit	ТФС	up to 100 μm
Carbonate taurit	ТК	0–35
Carbonate taurit (after desintegrator)	ТКД	up to 100 μm
Flotational shale taurit	ТФК	up to 100 μm

The chemical condition of shale and carbonate taurits is provided in Table 4.5 and 4.6.

Table. 4.5. **Chemical composition of shale taurits, % from bulk**

C	Zn	Cu	Ba	Mn	TiO_2	CaO	SiO_2	Fe_2O_3	Al_2O_3	K_2O	N_2O	LOI
10	0.01	0.00	0.04	0.09	0.50	4.00	68.0	2.88	5.01	0.40	0.40	8.67

Table. 4.6. **Chemical composition of carbonate taurits, % from bulk**

C	Zn	Cu	Ba	Mn	TiO_2	CaO	SiO_2	Fe_2O_3	Al_2O_3	K_2O	N_2O	LOI
12	0.01	0.01	0.05	0.09	0.5	31.0	38.00	2.69	4.55	1.20	0.40	9.51

Koksu Mining Company LLC produces six grades of taurits with their intended use (Table 4.7).

Enterprises-consumers of taurit products are similar to those enterprises using Karelian shungite products.

Table. 4.7. **Taurit trademarks and their intended use**

	Grade	Technical conditions	Intended use
Shale taurit	TC	TU 19 00 RK-39646043-003-2003 Granulated substance ranged from taupe to black	As a natural mineral drying agent: – for treating, fining, neutralization, bioactivation of natural water for drinking and household needs; – for distilling production and adsorption of some negative and positive ions from water media in filtering units; – for treating and neutralizing components in waste and circulating water, in particular clearing from oil products, phenols, mercury, heavy metals and chlorine
Carbonate taurit	TK	– // –	– // –
Shale taurit	ТСД	TU 19 00 RK-39646043-003-2003 Finely-dispersed powder ranged from taupe to black (depending on the size distribution)	For producing: – household chemicals as mineral pigments and fillers; – polymer filled and highly filled composition materials of multi-purpose nature; – rubber mixes for producing different industrial-rubber goods as fillers, modifiers or intermolecular plasticizer; – acid-proof filler for making acid-proof plasters, resins, matrices and concretes for protecting equipment, civil structures and gas flue
Shale taurit	ТСД	TU 19 00 RK-39646043-003-2003 Finely-dispersed powder ranged from taupe to black (depending on the size distribution)	from inorganic acids, salts, acidic gases; – for producing structural elements, structures and structural products, damp-proof rolled roofing and other products; – as a filler, stable natural mineral black pigment; – for protection from gamma photons of different energy; – as physiotherapeutic medical pastes and admixtures

	Grade	Technical conditions	Intended use
Carbonate taurit	ТКД	TU 19 00 RK-39646043-003-2003 Finely-dispersed powder ranged from taupe to black (depending on the size distribution)	For absorbing mercury vapor during manufacturing structural elements, structures and structural products, damp-proof rolled roofing, etc. – as a fertilizer; – for production of physiotherapeutic medical pastes and admixtures.
Flota-tional shale taurit	ТФС	TU 19 00 RK-39646043003-2003 Finely-dispersed powder ranged from taupe to black (depending on the size distribution)	For production of polymer filled and highly filled composition materials of multi-purpose nature; – for producing rubber mixes for producing different industrial-rubber goods as fillers, modifiers or intermolecular plasticizer; – for production of fire-resistant components in parting paints and mold materials; – for polymetallic and gold-containing ore concentration, sorption of reagents from pulp.
Flota-tional carbonate taurit	ТФК	TU 19 00 RK-39646043-003-2003 Finely-dispersed powder ranged from taupe to black (depending on the size distribution)	For manufacturing structural elements, structures and structural products, damp-proof rolled roofing, etc. – for producing rubber mixes for producing different industrial-rubber goods as fillers, modifiers or intermolecular plasticizer;

Products from shungite rocks mined in Karelia and Kazakhstani tau-rits are widely used not only in Russia and Kazakhstan, but also in Germany, Poland, Ukraine, Moldova, Czech Republic, Estonia, Japan. The export of shungite products increases each year (Table 4.8)

Table. 4.8. Production and consumption of shungite products in Russia
(in thousand tons)

Value	Years					
	2002	2003	2004	2005	2006	2007
Production	58.5	79.1	81.8	170.9	287.8	310.1
Export	0	1	13.3	0.6	0.1	0.4
Import	0	0	0	0.08	0.27	0.15
Consumption	58.5	78.1	68.5	170.38	287.97	309.85

So shungite products have a multi-purpose nature and are widely used in different fields. In the future the scope of this products will be extended.

Conclusions

1. Detailed analysis of the composition and characteristics of the main components of shungite rock and the data obtained from practical use in several fields have allowed developing the process of obtaining products from shungite rock considering the coarseness category and chemical peculiarities of all types.

2. Comprehensive analysis of all types of shungite rock has shown that each type has its intended use. E.g., sodium silicate rocks can be used for producing liquid glass after processing in the steam chamber, aluminosilicate rock serves as good raw materials for producing haydite while as carbonate types of low-grade rocks are a source of lime after burning. The mixture of aluminosilicate and carbonate rocks is the best raw material for ordinary cement which is currently sought-after.

3. After mixing with carbonate rocks with a low content of shungite high-grade shungite rocks, which contain more than 65 % of carbonacous matter, can serve as a raw material for producing calcium carbide widely used for metal cutting and welding. Carbonate rocks with a low content of shungite can serve as a raw material for calcium chloride and calcium metal.

4. It was shown that burning turned acidic rocks with a low content of shungite and high content of calcium carbonate into alkaline ones. After quenching the calcinated product it is turned into lime which can be used in different fields. The technological properties of such rocks may differ significantly from those of non-calcinated acidic ones. Carbonate rocks with a low content of shungite can also be successfully leached with chlorohydric acid. The formed solution of calcium chloride formed can be separated from the solid residue by

filtering it in a flash drier for obtaining the powder of calcium chloride. Calcium metal can be obtained from the solution using electrolytic process. It is a sought-after product for rare metal industry.

5. It is assumed that the industrial manufacturing will significantly extend the intended use of several types of shungitous rocks. This will allow solving the issue of comprehensive use of all well-known types at the territory of Zazhogino deposit and other Karelian regions. It should be noted that low-grade shungite rocks are very complex and diverse. So, for solving tasks for best use of different types, works for additional mineral processing are required for getting the necessary raw materials for manufacturing new products. Such mineral processing of low-grade shungite rocks is performed by thermal and hydrometallurgical methods while in each certain case the mineral processing method depends on the chemical composition of original rocks and specifications of the customer.

6. Two large Russian enterprises manufacture shungite products: Carbon-Shungite RPC LLC and Kondopoga shungite plant LLC. Both enterprises manufacture shungite products of different coarseness and mineral processing is not provided. The chemical composition of the obtained products depends only on the mineral composition of the rocks at the deposit. There is one large enterprise in Kazakhstan — Koksu Mining Company LLC. Flotational mineral processing of shungite rocks is provided there. Viable directions of processing shungite rocks — chemical leaching, thermal processing, processing in a steam chamber — are now on the stage of experimental-industrial development.

Conclusion

As a result of the performed complex researches of all types of high-grade and low-grade shungite rocks at Zazhogino deposit their unique physical-chemical characteristics have been determined so as their viability of applying them in different fields . From the compositional point of view shungite rock is a peculiar organic carbonaceous composite where high-dispersity silicate inclusions are distributed homogeneously. Besides that, the main matrix of the rock has a great content of different mixtures in form of aluminum silicates, carbonates, ferriferous minerals. The rock also contains a wide range of trace elements: vanadium, titanium, copper, zinc, lead, chrome, cobalt, molybdenum, etc. They serve as process catalyzers of various reactions. Depending on the sampling location at the deposit their chemical composition and content of various admixtures are rather different. This influences the general technological properties within using shungite products.

The matrix of shungite rock is based on the organic carbonaceous substance with rather specific properties and structure which is rather different from those of other carbonaceous rocks such as blacks, cokes and mineral carbon.

Long-term researches performed by All-Russian scientific-research institute, KarRC RAS Institute of Geology and Mekhanobr Institute allowed developing the standard operating procedure. The shungite rock processing enterprise has been developed and constructed on its basis. It provides a dry flow chart of processing shungite rocks with 25–40 % of carbonaceous matter and 45–70 % of silica. However, further researches has shown that the potential intended use of other types of shungite rocks — from those with high carbon content to low-grade ones — is much wider that covered by the design. A great diversity of low-grade shungite rocks requires considerable additional researches. It may happen that industrial manufacturing and involving rocks with high and low content of carbonaceous matter the number of potential consumers will increase; the scope of industrial use will extend and all technological properties of both high- and low-grade shungite rocks will be sought-after. According to the results of the researches, additional mineral processing operation are needed for performing new process tasks of industrial application of particularly low-grade shungite rocks to improve the qualitative indicators of the obtained products.

The performed researches has shown that shungite rocks of Zazhogi-no deposit are very viable for using without prior mineral processing. The average carbon concentration in shungite rocks — 25–45 % and that of silica — 45–70 %. In fact, it is a ready-made concentrated shungite product applied in many fields: construction, metallurgical, chemical. industrial-rubber.

In the centre of the deposit shungite rocks have a rather uniform chemical composition and the concentration of carbonaceous matter is 25–55 %. In rocks from separate deposit sections there are significant fluctuations of the main chemical components, i.e. carbonaceous matter and silica. Thus, the content of carbonaceous matter can range from 1 to 98 %. The content of other components of shungite rocks is also different at these section: pelitic minerals, feldspar, carbonates, etc. from 1 to 40 %. The rocks from these sections require additional researches for determining their intended use.

Low-grade shungite rocks have a lot of mineral admixtures. Their content depends on their location and chemical properties of shungite rock. The impact of these admixtures on the shungite substance has not been studied yet. Their role will be studied during industrial implementation of shungite rock in different fields.

After burning acidic low-grade shungite rocks with calcium carbonate are turned into alkaline ones. The technological properties of such rocks may differ significantly from those of non-calcinated acidic ones. However, due to lack of industrial experience their intended use and technological properties have not been determined yet. These properties can be determined only after industrial testing. The technological aspect of such researches on the carbonate shungite product is of great scientific and practical interest.

As a result of the analysis of structural, physical-chemical and mineralogical characteristics of all types of shungite rock it has been determined that they mostly contain organic carbonaceous matter with a hollow globular frame structure with distributed finely-dispersed silica and other mineral admixtures inside.

A special detailed analysis of the clean carbonaceous matter from different sections of the deposit has shown that it ubiquitously contains a little amount of sulphur dioxide, oxygen, nitrogen, hydrogen. From our point of view, these components are included into globular organic structure of the carbonaceous matter substituting the carbon atoms. The percentage of volatile matter is small and is 1.5–2.5 % in average. These chemical elements partially escape under ordinary temperature and when rock are heated to 100–150 °C they fully disappear. Presence of these components in the carbonaceous matter proves their organic origin. This largely determines the physical-chemical properties of the carbonaceous matter. Sulphur and oxygen play a special role in oxidation-reduction reactions. Due to carbonaceous matter of gaseous components inside it is impossible to obtain chemically clean carbon. Usually the maximum content of carbonaceous matter which can be extracted in form of concentrate does not exceed 98 %.

It was determined that besides the main carbonaceous matter the shungite rock also contains various modifications of finely-dispersed silica in form of disseminated shots, veinlets and noddles. Finely-dispersed silica has a rather diverse form and structure, the crystal coarseness ranges from 1 to 100 μm.

As a result of performed researches the mechanism of forming acid compounds in shungite rocks was determined. It is shown that the acidity of shungite rocks depends on the content of elemental sulphur in the carbonaceous matter and on sulphide and sulphate presence while the main component which causes the acid reaction is elemental sulphur. Sulphide and sulphate minerals present in the shungite rock oxidize slowly under ordinary temperature, however when drying shungite products they begin to decompose under 100–150 °C, especially during long-term storage. Under 400 °C a large part of sulphides is decomposed almost fully, intensidying the acid condition of the whole shungite rock. The process of decomposing sulphides can intensively run during drying; the formed sulphur dioxide is educed into air polluting the environment and causing significant harm to the sanitary-hygiene environment around the industrial facilities. For cleaning the air from sulphur-containing compound special procedures are required. It is recommended to perform such cleaning by breathing through water medium.

A detailed analysis of technological properties of main components of shungite rocks and the data obtained from their practical application in different fields allowed developing a basically new classification of shungite products according to coarseness categories and creating the best flow chart on the basis of this classification. It includes not only two stages of crushing, size specification of crushed products, electric drying, but also additional operations: burning, processing in the steam chamber, chemical leaching.

According to this classification it is necessary to develop new technical conditions for industrial application of all shungite products of various categories of coarseness. Within the process of size specification of products at the vibratory sieve the size of its mesh can be changed in order to obtain any coarseness categories considering the consumers' requirements.

It was shown that all-sliming of shungite rocks could be performed in the drum mill or in other grinding machines while it was more efficient to perform grinding in a jet-type mill working together with air classifier-separator, cyclones and sock filters. Depending on the working mode of the separator, cyclones and sock filters finely-dispersed shungite products of several coarseness categories can be obtained.

Based on the theoretical considerations and literature generalizations a procedure of drying shungite rocks using differential scanning calorimetry has been developed. The temperature parameters was set and the drying process was substantiated. This allowed determining the best operating mode of work of drying units and substantiating the selection of the required equipment. It was shown that it was the best of all to dry shungite products in a ПЭВ-270 electrical vibrating drier which provides automatic control of technological parameters with minimal losses of energy.

The performed researches has shown that the jet-type mill is the best equipment for grinding shungite rocks. Depending on the industrial needs it provides any design coarseness required by the customer. Within operation the ground shungite product is not clogged by foreign (grated) ferriferous material. The mill is working with minimal noise within the allowable sanitary rules.

Using Fritsch® laser diffractive microanalyser the derived and integral curves of the size distribution of ground products sampled from different stages of the grinding cycle: air classifier, cyclone, sock filters; it has been substantiated that the jet-type mill provides the required coarseness of grinding in the narrow range of particle sizes avoiding product pulverization. It was shown that the products of a certain coarseness category required for industrial application could be obtained at each processing stage. Thus, the air classifier allows obtaining 50–70 μm products, cyclones 15–25 μm, sock filters — a 5–10 μm one. These products comply with technical conditions developed for industrial application.

Due to the fact that resurgent sulphur dioxide is formed during drying and all-sliming of shungite rocks the technology of their capture using flue gas washer and neutralizing has been developed. For this an internal closed-looped water rotation system shall be created at the constructed enterprise. A recommended flow chart of the internal closed-looped water rotation system allows performing a wet sanitary processing of workplaces and all process equipment on a shift basis. This allows reducing dust formation inside the shop.

As a result of the performed researches it was determined that rocks with low content of carbonaceous matter (0.5–15 %) can also be widely used in various fields including construction, mining and processing, chemical, metallurgical industry. It was established that all revealed types of low-grade shungite rocks — silicate, aluminosilicate, carbonate -silicate, sodium silicate, hybrid sinter-containing one, etc. — can be easily applied as original raw materials in different fields.

For clarifying the areas of application of shungite rocks a basically new classification of all revealed types according to the chemical composition and physical-chemical properties has been developed. It was shown that according to the content of carbonaceous matter all shungite rocks could be divided into two types: high- and low-grade ones. According to the new classification shungite rocks with high content of carbonaceous matter (more than 25 %) are divided into three types considering their intended use: high-shungite, medium-shungite and shungite. High-shungite rocks contain more than 65 % of carbonaceous matter. They shall be applied as a source of energy, i.e. fuel and as an artificial fertilizer. The second type — medium-shungite rocks with 45–65 % of carbonaceous matter — can be applied as raw materials and a filler for producing paints, insulants, in construction of walls with shielding properties in special-purpose institutions. It is also recommended to use these rocks as an additive to low-shungite rocks when mixing them for blending. Production of industrial-rubber is the main and most viable area of consumption of shungite rocks with 25–45 % of carbonaceous matter where synthetic and technical carbon (carbon black) and synthetic (reinforcing) silica are replaced with shungite products in producing tyres and elastomers. Besides that, this type of rock can be widely used in metallurgical industry and for producing construction materials, as proved by industrial practice.

Analysis of features and chemical composition of shungite rocks with low content of bicarbonates matter allowed classifying them considering their intended use into the following types: silicate, aluminosilicate, sodium silicate, aluminosilicate-carbonate, carbonate. Each type of low-grade shungite rocks has its intended use.

A triangle chart was developed on the basis of genetic peculiarities of Zazhogino deposit. It shows the potential interreactions between different types of shungite rock considering the content of carbonaceous matter, silicon oxides and other components: aluminium silicates, carbonates, etc. On the basis of the proposed chart a new technique of evaluating the viability of all types of shungite rock has been developed. Thus, for high-grade rocks only determination of the content of carbonaceous matter is required. Its presence allows determining the area of potential use. For evaluating low-grade rocks analysis of oxide content shall be performed, apart from determining the carbon content: silicon, aluminium, calcium, sodium. The type and intended use of rocks is determined on the basis of the content of these oxides and according the proposed chart.

Comprehensive analysis of all types of low-grade shungite rock has shown that each of them has its own intended use. E.g., sodium silicate rocks can be used for producing liquid glass after processing in the steam chamber, aluminosilicate rock serves as good raw materials for producing haydite while as carbonate types are a source of lime after burning. The mixture of aluminosilicate and carbonate rocks is the best raw material for ordinary cement which is currently sought-after.

It is assumed that the industrial manufacturing will significantly extend the intended use of several types of shungitious rocks. This will allow solving the problem of their complex use. But it should be underlined that the features and composition of shungitious rocks are very complex and diverse and that the adopted classification is rather conditional. So in some cases the original rock shall be subject to additional mineral processing for solving optimum problems of their usage. Such processing can be done either by dry or washing method. For performing such works additional researches are required for sampling the rocks with the best composition in every certain case.

The performed researches allow accelerating the implementation of shungite rocks into industrial practice and extending their intended use in different fields.

Literature

Awdochin W.M., Rafijenko W.A., Pawłowa E.P. Nowy kierunek w składowaniu odpadów w zakładach wzbogacania kopalin. Uniwersytet Kardynała Stefana Wyszyńskiego w Warszawie. Instytut Ecologii i Bioetyki. Studia Ecologiae et Bioethicae. Tom 3/2005. Warszawa: Wydawnictwo UKSW – 2006. P. 349–353.

Krauskopf K.B. Dissolution and precipitation of silica at low temperatures // Geochimica et Cosmochimica Acta. Elsevier, 1956. V. 10. P. 137–148.

Авдохин В.М., Абрамов А.А. Окисление сульфидных минералов в процессах обогащения. М.: Недра, 1989. 232 с.

Ануфриева С.И. Оценка изменения вещественного состава и физико-химических свойств шунгитового сорбционного материала при его модификации. Труды V-го конгресса обогатителей стран СНГ. М.: МИСиС, 2005. С. 143–145.

Атлас текстур и структур шунгитоносных пород Онежского синклинория / Под ред. М.М. Филиппова, В.А. Мележека. Петрозаводск: КарНЦ РАН, 2006. 80 с.

Бардовский А.Д., Жуков В.П., Перевалов В.С., Рафиенко В.А. Производство щебня из карбонатных пород с использованием шнековых грохотов // Горный информационно-аналитический бюллетень. 2003. № 9. С. 151–152.

Бардовский А.Д., Рафиенко В.А. Разработка грохотов с непосредственным возбуждением сита // Горный информационно-аналитический бюллетень, 2003. № 7. С. 17–23.

Банникова Л.А. Органическое вещество в гидротермальном рудообразовании. М.: Наука, 1990. 207 с.

Болдырев А.К., Ковалёв Г.А. Рентгенометрические исследования шунгита, антрацита и каменного угля // Зап. ЛТИ. 1937. Т. 10. Вып. 2. 12 с.

Бондарь Е.Б., Клесмент И.Р., Куузик М.Г. Исследование структуры и генезиса шунгита // Горючие сланцы. 1987. № 4. С. 377–393.

Борисов П.А. Карельские шунгиты. Петрозаводск: Госиздат КаССР, 1956. 92 с.

Борисова Р.И., Климов Н.И. Отчёт о результатах ревизионно-опробовательских работ на вспучивающиеся сланцы и о предварительной разведке Красносельского месторождения шунгитсодержащих пород. Петрозаводск: ККГРЭ, 1974. 210 с.

Боровский И.Б., Блохин М. Рентгеноанализ карельского шунгита. Отчёт Механобра. 1933 // Фонды КарНЦ РАН. Ф. 1, оп. 24, ед. хр. 71.

Буллах А.Г., Абакумова Н.Б. Каменное убранство центра Ленинграда. Л.: Изд-во ЛГУ, 1987. 286 с.

Вассоевич Н.Б. Основные закономерности, характеризующие органическое вещество современных и ископаемых осадков. В кн.: Природа органического вещества современных и ископаемых осадков. М.: Наука, 1973. С. 11–59.

Вильшанский А.И., Маслаков В.А., Рафиенко В.А. Оборудование для тонкого измельчения минерального и органоминерального сырья // Труды I-го Международного форума «Рациональное природопользование». М.: ПИК МАКСИМА, 2005. С. 214–215.

Виноградов А.П. Распространённость элементов в горных породах. Краткий справочник по геохимии / Под ред. Г.В. Войткевича и др. М.: Наука, 1977. 77 с.

Винокуров С.Ф., Новиков Ю.Н., Усатов А.В. Фуллерены в геохимии эндогенных процессов // Геохимия. 1997. № 9. С. 937–944.

Войткевич Г.В., Мирошников А.Е., Поваренных А.С. и др. Краткий справочник по геохимии. М.: Недра, 1987. 183 с.

Волков И.И. Геохимия серы в осадках океана. М.: Наука, 1984. 271 с.

Галдобина Л.П. Металлогения шунгитсодержащих и шунгитовых пород Онежской мульды // Материалы по металлогении Карелии. Петрозаводск, 1987. С. 100–113.

Галдобина Л.П., Горлов В.И. Фациально-циклический анализ шунгитсодержащих толщ заонежской свиты (верхний ятулий) Карелии. В кн.: Геология и полезные ископаемые Карелии. Петрозаводск, 1975. С. 103–109.

Галдобина Л.П., Калинин Ю.К., Купряков С.В. Эндогенное происхождение шунгитовых пород протерозоя Карелии // Труды 2-го Всесоюзного совещания по геохимии углерода: Тезисы докладов. М.: Наука, 1986. С. 79–81.

Галкин С. Промышленное освоение шунгита и планы работ на дальнейшее исследование // Труды 2-й Карельской геологоразведочной конференции. Петрозаводск, 1933. С. 35–41.

Глебашев С.Г. Минеральное сырьё. Шунгит. М.: Геоинформмарк. 1999. 17 с.

Головенок В.К. Докембрийские микрофоссилии в кремнях и их биостратиграфическое значение // Сов. геология. 1989. № 8. С. 41–48.

Голубев А.И., Ахмедов А.М., Галдобина Л.П. Геохимия черносланцевых комплексов нижнего протерозоя Карело-Кольского региона. Л.: Изд-во ЛГУ, 1984. 192 с.

Горлов В.И. Генезис шунгита // Шунгитовые сланцы Карелии — новый вид сырья для производства эффективных строительных материалов. Отчёт по т. 10. Т. 1. Петрозаводск, 1966 (Фонды КарНЦ РАН. Ф. 13, оп. 5, ед. хр. 72).

Горлов В.И., Калинин Ю.К., Иванова И.Е. Разработка технологии и геологическое изучение шунгитовых пород как комплексного сырья. Отчёт по т. 30 // Фонды КарНЦ РАН. Петрозаводск, 1977.

Горлов В.И., Калинин Ю.К., Костинюк Г.П. и др. Опробование и изучение Нигозёрских сланцев как сырья для производства лёгких пористых заполнителей. Отчёт. Петрозаводск, 1962 (Фонды КарНЦ РАН).

Горштейн А.Е., Барон Н.Ю., Сыркина М.Л. Адсорбционные свойства шунгитов // Известия ВУЗов, химия и химическая технология 1979. Т. 22. № 6. С. 711–715.

Доронина Ю.А. Шунгит — камень-спаситель. СПб.: Невский проспект, 2004. 96 с. Серия «Целительные силы природы».

Дюккиев Е.Ф. Пористая структура и удельная поверхность // Шунгиты — новое углеродистое сырьё. Петрозаводск, 1984. С. 105–106.

Дюккиев Е.Ф., Калинин Ю.К., Туполев А.Г. Основные физические и физико-химические характеристики шунгитов Зажогинского карьера // Технологические свойства и характеристики минерального сырья Карелии. Петрозаводск, 1986. С. 28–32.

Елецкий А.В., Смирнов Б.М. Фуллерены и структуры углерода // Успехи физических наук. 1995. Т. 165. № 9. С. 977–1009.

Жумалиева К. Рентгенографическое исследование структуры и термических преобразований шунгита: Автореф. дисс.... канд. геол.-мин. наук. Симферополь: Изд-во СГУ им. Фрунзе, 1974. 15 С.

Зайденберг А.З., Ковалевский В.В., Рожкова Н.Н. и др. О фуллереноподобных структурах шунгитового углерода // Журнал физической химии. 1996. Т. 70. № 1. С. 107–110.

Зискинд М.С. Декоративно-облицовочные камни. Л.: Недра, 1989. 255 с.

Иванкин П.Ф., Галдобина Л.П., Калинин Ю.К. Шунгиты: проблемы генезиса и классификации нового вида углеродистого сырья // Современная геология. 1987. № 12. С. 40–47.

Иванова В.П., Касатов Б.К., Красавина Т.Н. и др. Термический анализ минералов и горных пород. Л.: Недра, 1974. 399 с.

Иностранцев А.А. Новый крайний член в ряду аморфного углерода // Горный журнал. 1879. Т. 11. № 5–6. С. 314–342.

Иностранцев А.А. О происхождении шунгита // Труды СПб общества естествоиспытателей. 1914. Т. 35. Вып. 5.

Иностранцев А.А. О шунгите // Горный журнал. 1886. № 2. С. 35–45.

Калинин Ю.К. Классификация шунгитовых пород // Шунгиты — новое углеродистое сырьё. Петрозаводск, 1984. С. 4–16.

Калинин Ю.К. Углеродсодержащие шунгитовые породы и их практическое использование: Автореф. дисс. … докт. техн. наук. М., 2002. 50 с.

Калинин Ю.К. Шунгитсодержащие породы в производстве шунгизита. В кн.: Шунгиты Карелии и пути их комплексного использования. Петрозаводск, 1975. С. 110–140.

Калинин Ю.К., Горлов В.И. Вещественный состав шунгитового вещества // Шунгиты Карелии и пути их комплексного использования. Петрозаводск, 1975. С. 44–55.

Калинин Ю.К., Горлов В.И. Сырьё для производства чёрного пегмента // Строительные материалы. 1968. № 7.

Калинин Ю.К., Каляда Т.В., Соловов В.К. Строительные материалы на основе шунгитовых пород — эффективное средство биологической защиты населения от электромагнитных полей // Геология и охрана недр Карелии. Петрозаводск, 1992. С. 51–60.

Калинин Ю.К., Ковалевский В.В. Электронно-микроскопическое исследование структуры шунгитов // Минеральное сырьё Карелии. Петрозаводск, 1977. С. 119–124.

Калинин Ю.К., Пеки А.С. Шунгитовые породы как адсорбент // Минеральное сырьё Карелии. Петрозаводск, 1977. С. 212.

Калинин Ю.К., Филиппов М.М., Капутин Ю.Е., Мутыгуллин Р.Х. Качество и эффективность использования шунгизитового сырья Карелии. Петрозаводск, 1988. 147 с.

Калмыков Г.С. Свойства метаморфизованного сапропелита (на примере Карельского шунгита) // Проблемы геологии нефти. 1974. Вып. 4. С. 264–274.

Касаточкин В.И., Элизен В.М., Мельниченко В.М. и др. Субмикропористая структура шунгита // Химия твёрдого топлива. 1978. № 3. С. 17–21.

Кастальский А.А. Проектирование устройств для удаления из воды растворённых газов в процессе водоподготовки. М.: Госстройиздат, 1957. 148 с.

Кашкина Л.В., Кашкин В.Б., Рублева Т.В., Шикунова О.А. Изучение физических свойств фуллеренов и фуллерено-содержащих саж: Учебное пособие / Красноярский образовательный центр высоких технологий. Красноярск: Изд-во СAA им. М.Ф. Решетнёва, 2000. 80 с.

Ковалевский В.В. Электронно-графическое исследование шунгитов: Автореф. дис. … канд. физ.-мат. наук. М., 1986. 17 с.

Ковалевский В.В. Шунгитовые породы Карелии — современные представления о строении и перспективах использования в наукоемких технологиях // Северная Европа в XXI веке: природа, культура, экономика. Т. 1. Материалы Международной конференции, посвященной 60-летию КарНЦ РАН (24–27 октября 2006 г.). Секция «Биологические науки». Секция «Науки о Земле». Петрозаводск: КарНЦ РАН, 2006. С. 271–273.

Коневский М.Р., Минин В.И. Шунгиты как комплексный сырьевой материал фосфорного производства // Шунгиты Карелии и пути их комплексного использования. Петрозаводск, 1975. С. 150–167.

Королёв Ю.М. Рентгенографическое исследование органического вещества саприпелевого типа // Геология нефти и газа. 1989. № 9. С. 50–53.

Крыжановский В.И. Геохимия месторождений шунгита // Минеральное сырьё. М.: ВИМС, 1931. № 10–11. С. 955–968.

Крыжановский В.И. Шунгит в Карелии // Советская Карелия. 1931. № 8/10. С. 15–19.

Крылов И.О., Луговская И.Г., Соколова В.Н., Ануфриева С.И. Материалы Международного совещания: Направленное изменение физико-химических свойств минералов в процессах обогащения полезных ископаемых (Плаксинские чтения – 2003). Петрозаводск, 2003. С. 128.

Купряков С.В. Геология и генезис шунгитовых пород Зажогинского месторождения // Органическое вещество шунгитоносных пород Карелии. Петрозаводск, 1994. С. 93–98.

Левин А.С. Основные вопросы геологии месторождений горючих сланцев. М.: Недра, 1982. 78 с.

Леманов В.В., Балашова Е.В., Шерман А.Б. и др. Акустические свойства шунгитов // ФТТ. Т. 35. № 11. С. 3082–3086.

Маслаков В.А., Исаев В.И., Вильшанский А.И., Рафиенко В.А. Комплексная технология производства дисперсного шунгита (Экологические аспекты) // Труды I-го Международного форума «Рациональное природопользование» М.: ПИК МАКСИМА, 2005. С. 216–217.

Маслаков В.А., Исаев В.И., Ануфриева С.И. и др. Проведение минералого-аналитических исследований природных типов шунгитовых пород и продуктов их технологической переработки с целью прогнозирования их потребительских свойств: Итоговый научно-технический отчёт. М.: МКК-Инжиниринг, ФГУП «ВИМС им. Н.М. Федоровского», 249 с.

Матвеев М.А. Растворимость стеклообразных силикатов натрия. М.: Промстройиздат, 1957. 92 с.

Мележик В.А. Седиментационные и органо-породные бассейны раннего протерозоя Балтийского щита: Автореф. дисс. … докт. геол.-мин. наук. Апатиты, 1987.

Мелентьев Б.Н., Иваненко В.В., Памфилова Л.А. Растворимость некоторых рудообразующих сульфидов в гидротермальных условиях. М.: Наука, 1968. 103 с.

Методы определения диоксида углерода карбонатов. ГОСТ 13455-91. Топливо твёрдое минеральное. М.: ИПК Издательство стандартов, 1992.

Методы определения зольности. ГОСТ 11022-90. Топливо твёрдое минеральное. М.: ИПК Издательство стандартов, 1991.

Методы определения углерода и водорода. ГОСТ 2408.1-88. Твёрдое топливо. М.: ИПК Издательство стандартов, 1989.

Мицюк Б.М., Горогонская Л.И. Физико-химические превращения кремнезёма в условиях метаморфизма. Киев: Наукова думка, 1980. 234 с.

Мишунина З.А. Литогенез органического вещества и первичная миграция нефти в карбонатных формациях. Л.: Изд-во ЛГУ, 1978. 152 с.

Муравьёв В.И. О природе глобулярного опала в опоках и трапелах. ДАН СССР. 1975. Т. 222. № 3. С. 684–686.

Нартов А.А. О пользе минералогии в отношении к хлебопашеству // Труды Вольного экономического общества. 1762. Ч. 2.

Новые месторождения ископаемого горючего в России // Горный журнал. 1877. Т. IV. № 10–12. С. 117–122.

Органическое вещество шунгитоносных пород Карелии (генезис, эволюция, методы изучения) / Под ред. М.М. Филиппова и А.И. Голубева. Петрозаводск, 1994. 208 с.

Орлов А.Д. Шунгит — камень чистой воды. СПб.: Диля, 2004. 112 с.

Отчёт по 5-му этапу работ по исследованию шунгитовых пород. М.: МКК-Инжиниринг, ФГУП «ВИМС им. Н.М. Федоровского», 2002.

Парфёнова Л.С., Волконская Т.И., Тихонов В.В. и др. Теплопроводность, теплоёмкость и термо-ЭДС шунгитового углерода // ФТТ. 1994. Т. 36. № 4. С. 1150–1153.

Пасманик М.И., Сасс-Тисовский Б.А., Якименко Л.М. Производство хлора и каустической соды: Справочник. М.: Химия, 1966. 312 с.

Пекки А.С. Использование шунгитовых пород в сельском хозяйстве // Шунгиты Карелии и пути их комплексного использования. Петрозаводск, 1975. С. 214–216.

Пеньков В.Ф. Генетическая минералогия углеродистых веществ. М.: Недра, 1996. 224 с.

Перельман А.И. Геохимия. М.: Высшая школа, 1979. 423 с.

Полеховский Ю.С., Резников В.А. Фуллерены — новое природное сырьё // Образование и локализация руд в земной коре. СПб.: Наука, 1999. С. 123–147.

Преображенский Н.А. Рентгеновская кристаллография и структура углеродистых веществ // Химия твёрдого топлива. 1992. № 5. С. 93–99.

Рафиенко В.А., Бардовский А.Д. Грохот с комбинационным возбуждением сита // Труды VII-ой Международной экологической конференции студентов и молодых учёных «Экологическая безопасность как ключевой фактор устойчивого развития». М.: Ойкумена, 2003. Т. 2. С. 89–91.

Рафиенко В.А., Зубков Д.Г. Механизм естественного кислотообразования шунгитовых пород и его роль при промышленном внедрении. М.: Изд-во НПП «Фильтроткани», 2019. 100 с.

Рафиенко В.А., Малюк О.П. Обогащение кварцевых песков. М.: Изд-во МГГУ, 2004. 55 с.

Рафиенко В.А. Разработка технологии внутреннего замкнутого водооборота при производстве шунгитовых продуктов. Труды V-го Конгресса обогатителей стран СНГ. М.: Изд-во МИСиС, 2005. Т. 1. С. 190–194.

Рафиенко В.А. О механизме выщелачивания сульфидов из шунгитовых пород // Горный информационно-аналитический бюллетень. 2007. № 9. С. 38–48.

Рафиенко В.А. Технология переработки шунгитовых пород. М.: Геос, 2008. 214 с.

Рафиенко В.А. К вопросу термического разложения сульфидов в шунгитовых породах // Горный информационно-аналитический бюллетень. 2007. № 8. С. 89–90.

Рафиенко В.А. Разработка математической модели процесса грохочения // Горный информационно-аналитический бюллетень. 2008. № 1. С. 53–56.

Рожкова Н.Н. Влияние шунгитового наполнителя на физико-механические и проводящие свойства полимерных композиционных материалов: Автореф. дисс. ... канд. техн. наук. Петрозаводск, 1992. 17 с.

Рожкова Н.Н. Механические характеристики // Шунгиты — новое углеродистое сырьё. Петрозаводск, 1984. С. 62–65.

Рожкова Н.Н., Андриевский Г.В. Фуллерены в шунгитовом углероде // Фуллерены и фуллереноподобные структуры. Минск, 2000. С. 63–69.

Рысьев О.А. Шунгит — вечный хранитель здоровья. СПб.: Диля, 2003. 192 с.

Рысьев О.А. Шунгит — камень жизни. СПб.: Диля, 2004. 128 с.

Рычанчик Д.В., Ромашкин А.Е. Особенности внутреннего строения Максовской залежи шунгитовых пород // Углеродсодержащие формации в геологической истории: Труды международного симпозиума. Петрозаводск, 2000. С. 73–79.

Рябов Н.И. Очерк шунгитовых месторождений Карелии. Петрозаводск, 1948. 51 с. (Фонды КПСЭ).

Рябов Н.И. Шунгиты Карелии // Труды 2-й Карельской геолого-разведочной конференции. Петрозаводск. 1933. С. 30–35.

Сиваев В.В. Строение шунгито-карбонатно-сланцевой толщи верхнего ятулия в северо-западном Прионежье // Геология и полезные ископаемые Карелии. Петрозаводск, 1975. С. 91–102.

Сиверцев А.П. Исследование свойств шунгита: Отчёт. 1957. // Фонды КарНЦ РАН. Ф. 4, оп. 5, ед. хр. 58. 99 с.

Симонейт Б.Р.Т. Созревание органического вещества и образование нефти: гидротермальный аспект // Геохимия. 1986. № 2. С. 236–254.

Соколов В.А., Калинин Ю.К. Тектонические и практические аспекты проблемы шунгитов // Вестник АН СССР. 1976. № 5. С. 76–84.

Стащук М.Ф. Проблема окислительно-восстановительного потенциала в геологии. М.: Недра, 1968. 208 с.

Технологический регламент опытно-промышленного участка по производстве шунгитовых продуктов. М.: МКК-Инжиниринг, ФГУП «ВИМС им. Н.М. Федоровского», 2004. 43 с.

Тойкка М.А. Шунгит как местное удобрение // Уч. зап. КФГУ. Т. 1. Петрозаводск, 1946. С. 215–269.

Туполев А.Г., Дюккиев Е.Ф. Теплопроводность. Теплоёмкость // Шунгиты — новое углеродистое сырьё. Петрозаводск, 1984. С. 82–84.

Тяганова В.И., Калинин Ю.К. Термическое расширение. Термическая стойкость // Шунгиты — новое углеродистое сырьё. Петрозаводск, 1984. С. 71–77.

Угли бурые, каменные, антрацит, сланцы горючие и торф. Ускоренный метод определения углерода и водорода. ГОСТ 6389-81. М., 1981. 12 с.

Федосеев С.Д. Неизотермичность гетерогенных реакций и проблема газификации твёрдого топлива: Автореф. дисс. … докт. техн. наук. М.: Изд-во МХТИ, 1963. 45 с.

Филиппов М.М. Шунгитоносные породы Карелии: чёрная Олонецкая земля, аспидный сланец, антрацит, шунгит. Петрозаводск: КарНЦ РАН, 2004. 488 с.

Филиппов М.М. Исходное органическое вещество шунгитовых пород Карелии // Очерки геологии докембрия Карелии. Петрозаводск: Рос. АН Карел. науч. центр. Ин-т геологии, 1995. С. 33–51.

Филиппов М.М. Модели формирования месторождений шунгитоносных пород Онежского синклинория: Автореф. дисс. … докт. геол.-мин. наук. Петрозаводск, 2000. 56 с.

Филиппов М.М. Ядерно-геофизические методы определения свободного углерода шунгитовых пород (постановка задачи) // Проблемы изучения докембрийских образований геофизическими методами. Петрозаводск, 1990. С. 40–57.

Филиппов М.М., Горлов В.И. Методика изучения месторождений шунгитсодержащих пород: Отчёт по договору о передаче науч.-техн. разработки. Петрозаводск, 1987. 46 с. (Фонды КарНЦ РАН).

Филиппов М.М., Калинин Ю.К., Горлов В.И. и др. Способ разведки месторождений полезных ископаемых. Авт. свид. № 915052, СССР, МКИ G01 9/00. Опубл. 23.03.82. Бюл. № 11.

Филиппов М.М., Кевлевич В.И. Проблема получения концентратов шунгитового вещества: Труды V-го Конгресса обогатителей стран СНГ. М.: Изд-во МИСиС, 2005. С. 54–57.

Филиппов М.М., Медведев П.В., Ромашкин А.Е. Метаколлоидная природа шунгитовых пород // Углеродсодержащие формации в геологической истории: Тезисы докладов международного симпозиума. Петрозаводск, 1998. С. 60–61.

Филиппов М.М., Медведев П.В., Ромашкин А.Е. О природе шунгитов Южной Карелии // Литология и полезные ископаемые. 1998. № 3. С. 323–332.

Филиппов М.М., Ромашкин А.Е. Шунгитовые породы — генезис, классификация, методы определения $C_{св}$. Петрозаводск, 1996. 90 с.

Филиппов М.М., Савицкий А.И., Соколов С.Я. Способ разведки месторождений полезных ископаемых. Авт. свид. № 1166043, СССР, МКИ G01 9/00. Заявлено 03.02.84. Опубл. 07.07.85. Бюл. № 825.

Фирсова С.О., Шатский Г.В. Брекчии в шунгитовых породах Карелии и особенности их генезиса // ДАН СССР. 1988. Т. 302. С. 177–180.

Хант Дж. Геохимия и геология нефти и газа. М.: Мир, 1982. 704 с.

Хворова И.В., Дмитрик А.П. Микроструктуры кремнистых пород. М.: Наука, 1972. 253 с.

Холодкевич С.В., Бекренев А.В., Донченко В.А. и др. Экстракция природных фуллеренов из Карельских шунгитов // Доклады РАН. 1993. Т. 330. № 3. С. 340–342.

Чантурия В.А. Основные направления комплексной переработки минерального сырья // Горный журнал. 1995. № 1. С. 50–54.

Чекалова П.М. Минерало-петрографическое исследование Карельского шунгита: Отчёт Механобра. 1932. (Фонды Кар НЦ РАН).

Чухров Ф.В. Коллоиды в земной коре. М.: Изд-во АН СССР, 1955.

Шунгитовые породы для производства электропроводящих строительных материалов. ТУ 21-РСФСР-768-79. МПСМ РСФСР. 1982.

Шунгитовые фильтры (краткий каталог изделий фирмы ООО «Фильтры ММ», «Минеральная продукция»). СПб., 2000. 43 с.

Шунгиты — новое углеродистое сырьё. Петрозаводск: Карелия, 1984. 184 с.

Юдович Я.Э., Кетрис М.П., Мерц А.В. Элементы-примеси в чёрных сланцах. Екатеринбург: Наука, 1994. 304 с.

Юшкин Н.П. Глобулярная надмолекулярная структура шунгита: данные растровой туннельной микроскопии // Доклады РАН. 1994. Т. 337. № 6. С. 800–803.

Ягодкина Т.К. Десульфурация углей в процессах в процессах обогащения в СССР и за рубежом: Обзорная информация. Вып. 9. М.: ЦНИЭИуголь, 1984. 71 с.

CONTENTS

www.ingramcontent.com/pod-product-compliance
Lightning Source LLC
Chambersburg PA
CBHW050730030426
42336CB00012B/1491